# 2016 SQA Past Papers & Hodder Gibson Model Papers With Answers

## Advanced Higher

# PHYSICS

SQA Advanced Higher PHYSICS

2015 Specimen Question Paper,
Model Papers and 2016 Exam

This book contains the official 2015 SQA Specimen Question Paper and 2016 Exam Paper for Advanced Higher Physics, with associated SQA-approved answers modified from the official marking instructions that accompany the paper.

In addition the book contains model papers, together with answers, plus study skills advice. These papers, some of which may include a limited number of previously published SQA questions, have been specially commissioned by Hodder Gibson, and have been written by experienced senior teachers and examiners in line with the new Advanced Higher for CfE syllabus and assessment outlines. This is not SQA material but has been devised to provide further practice for Advanced Higher examinations in 2016 and beyond.

Hodder Gibson is grateful to the copyright holders, as credited on the final page of the Answer Section, for permission to use their material. Every effort has been made to trace the copyright holders and to obtain their permission for the use of copyright material. Hodder Gibson will be happy to receive information allowing us to rectify any error or omission in future editions.

Hachette UK's policy is to use papers that are natural, renewable and recyclable products and made from wood grown in sustainable forests. The logging and manufacturing processes are expected to conform to the environmental regulations of the country of origin.

Orders: please contact Bookpoint Ltd, 130 Park Drive, Milton Park, Abingdon, Oxon OX14 4SE. Telephone: (44) 01235 827720. Fax: (44) 01235 400454. Lines are open 9.00–5.00, Monday to Saturday, with a 24-hour message answering service. Visit our website at www.hoddereducation.co.uk. Hodder Gibson can be contacted direct on: Tel: 0141 333 4650; Fax: 0141 404 8188; email: hoddergibson@hodder.co.uk

This collection first published in 2016 by
Hodder Gibson, an imprint of Hodder Education,
An Hachette UK Company
211 St Vincent Street
Glasgow G2 5QY



Typeset by Aptara, Inc.

Printed in the UK

A catalogue record for this title is available from the British Library

ISBN: 978-1-4718-9080-2

3 2 1

2017 2016

# Introduction

## Study Skills – what you need to know to pass exams!

### Pause for thought

Many students might skip quickly through a page like this. After all, we all know how to revise. Do you really though?

*Think about this:*

"IF YOU ALWAYS DO WHAT YOU ALWAYS DO, YOU WILL ALWAYS GET WHAT YOU HAVE ALWAYS GOT."

Do you like the grades you get? Do you want to do better? If you get full marks in your assessment, then that's great! Change nothing! This section is just to help you get that little bit better than you already are.

There are two main parts to the advice on offer here. The first part highlights fairly obvious things but which are also very important. The second part makes suggestions about revision that you might not have thought about but which WILL help you.

### Part 1

DOH! It's so obvious but …

*Start revising in good time*

Don't leave it until the last minute – this will make you panic.

Make a revision timetable that sets out work time AND play time.

*Sleep and eat!*

Obvious really, and very helpful. Avoid arguments or stressful things too – even games that wind you up. You need to be fit, awake and focused!

*Know your place!*

Make sure you know exactly **WHEN and WHERE** your exams are.

*Know your enemy!*

**Make sure you know what to expect in the exam.**

How is the paper structured?

How much time is there for each question?

What types of question are involved?

Which topics seem to come up time and time again?

Which topics are your strongest and which are your weakest?

Are all topics compulsory or are there choices?

*Learn by DOING!*

There is no substitute for past papers and practice papers – they are simply essential! Tackling this collection of papers and answers is exactly the right thing to be doing as your exams approach.

### Part 2

People learn in different ways. Some like low light, some bright. Some like early morning, some like evening or night. Some prefer warm, some prefer cold. But everyone uses their BRAIN and the brain works when it is active. Passive learning – sitting gazing at notes – is the most INEFFICIENT way to learn anything. Below you will find tips and ideas for making your revision more effective and maybe even more enjoyable. What follows gets your brain active, and active learning works!

*Activity 1 – Stop and review*

Step 1

When you have done no more than 5 minutes of revision reading STOP!

Step 2

Write a heading in your own words which sums up the topic you have been revising.

Step 3

Write a summary of what you have revised in no more than two sentences. Don't fool yourself by saying, "I know it, but I cannot put it into words". That just means you don't know it well enough. If you cannot write your summary, revise that section again, knowing that you must write a summary at the end of it. Many of you will have notebooks full of blue/black ink writing. Many of the pages will not be especially attractive or memorable so try to liven them up a bit with colour as you are reviewing and rewriting. **This is a great memory aid, and memory is the most important thing.**

### *Activity 2 – Use technology!*

Why should everything be written down? Have you thought about "mental" maps, diagrams, cartoons and colour to help you learn? And rather than write down notes, why not record your revision material?

What about having a text message revision session with friends? Keep in touch with them to find out how and what they are revising and share ideas and questions.

Why not make a video diary where you tell the camera what you are doing, what you think you have learned and what you still have to do? No one has to see or hear it, but the process of having to organise your thoughts in a formal way to explain something is a very important learning practice.

Be sure to make use of electronic files. You could begin to summarise your class notes. Your typing might be slow, but it will get faster and the typed notes will be easier to read than the scribbles in your class notes. Try to add different fonts and colours to make your work stand out. You can easily Google relevant pictures, cartoons and diagrams which you can copy and paste to make your work more attractive and **MEMORABLE**.

### *Activity 3 – This is it. Do this and you will know lots!*

#### Step 1

In this task you must be very honest with yourself! Find the SQA syllabus for your subject (www.sqa.org.uk). Look at how it is broken down into main topics called MANDATORY knowledge. That means stuff you MUST know.

#### Step 2

BEFORE you do ANY revision on this topic, write a list of everything that you already know about the subject. It might be quite a long list but you only need to write it once. It shows you all the information that is already in your long-term memory so you know what parts you do not need to revise!

#### Step 3

Pick a chapter or section from your book or revision notes. Choose a fairly large section or a whole chapter to get the most out of this activity.

With a buddy, use Skype, Facetime, Twitter or any other communication you have, to play the game "If this is the answer, what is the question?". For example, if you are revising Geography and the answer you provide is "meander", your buddy would have to make up a question like "What is the word that describes a feature of a river where it flows slowly and bends often from side to side?".

Make up 10 "answers" based on the content of the chapter or section you are using. Give this to your buddy to solve while you solve theirs.

#### Step 4

Construct a wordsearch of at least 10 × 10 squares. You can make it as big as you like but keep it realistic. Work together with a group of friends. Many apps allow you to make wordsearch puzzles online. The words and phrases can go in any direction and phrases can be split. Your puzzle must only contain facts linked to the topic you are revising. Your task is to find 10 bits of information to hide in your puzzle, but you must not repeat information that you used in Step 3. DO NOT show where the words are. Fill up empty squares with random letters. Remember to keep a note of where your answers are hidden but do not show your friends. When you have a complete puzzle, exchange it with a friend to solve each other's puzzle.

#### Step 5

Now make up 10 questions (not "answers" this time) based on the same chapter used in the previous two tasks. Again, you must find NEW information that you have not yet used. Now it's getting hard to find that new information! Again, give your questions to a friend to answer.

#### Step 6

As you have been doing the puzzles, your brain has been actively searching for new information. Now write a NEW LIST that contains only the new information you have discovered when doing the puzzles. Your new list is the one to look at repeatedly for short bursts over the next few days. Try to remember more and more of it without looking at it. After a few days, you should be able to add words from your second list to your first list as you increase the information in your long-term memory.

## FINALLY! Be inspired...

Make a list of different revision ideas and beside each one write **THINGS I HAVE** tried, **THINGS I WILL** try and **THINGS I MIGHT** try. Don't be scared of trying something new.

And remember – "FAIL TO PREPARE AND PREPARE TO FAIL!"

# Advanced Higher Physics

## The course

The course comprises two whole and two half units:

- Rotational Motion and Astrophysics (1 unit)
- Quanta and Waves (1 unit)
- Electromagnetism (0·5 unit)
- Investigating Physics (0·5 unit)

### *Course assessment*

In order to gain an award in the course, you must pass the unit assessment in each topic. The first three topics will be internally assessed by your teacher/lecturer and the evidence externally verified. The assessments will be in the form of class tests.

### *Investigating Physics*

The Investigating Physics unit assessment will involve you planning your experiments and taking note of your experimental readings in your record of work/diary.

Ensure this record is neat and tidy so that when you come to type your report, it is easy to decipher. Your teacher will overview and sign your record of work/diary.

**This record could be externally verified by SQA and should be submitted to your teacher along with your project write up.**

## The examination

This will last 2·5 hours and will be marked out of **140 marks** and then scaled back to **100 marks**.

### *The project*

The report will be submitted to SQA for marking and will be marked out of **30 marks**.

This will then be added to the examination mark giving a total out of **130 marks**.

Experimental write ups should be typed up as you progress through the course. This will build up your skills and also allow you to easily go back and correct any mistakes or take on any suggestions from your teacher.

Become confident in the use of software packages that allow you to plot graphs including error bars. Find out how to find the gradient of a straight line and its associated uncertainty – the use of Linest function in the Microsoft package. This will save you an enormous amount of time!

See the SQA website for advice to candidates in relation to the project.

Advanced Higher Physics is an excellent qualification and although it is a step above Higher, the fact that you are in the class proves you have the ability to do well. It just takes hard work and application – the danger is to have too many distractions in your sixth year. Try to choose a project that suits and **stick to deadlines**!

## Examination tips

### *Mark allocation*

- **4 or 5 marks** will generally involve more than one step or several points of coverage.
- **3 marks** will generally involve the use of just one equation.

### *Standard 3-mark example*

Calculate the acceleration of a mass of 5 kg when acted on by a resultant of force of 10 N.

| Solution 1 | Solution 2 | Solution 3 |
|---|---|---|
| F = ma **(1)** | F = ma **(1)** | F = ma **(1)** |
| 10 = 5 a **(1)** | $a = \frac{m}{F} = \frac{5}{10}$ **(0)** | 10 = 5a **(1)** |
| $a = 2\,ms^{-2}$ **(1)** | $= 0{\cdot}5\,ms^{-2}$ | $a = 0{\cdot}5\,ms^{-2}$ |
| **3 marks** | **1 mark for selecting formula.** | **2 marks for selecting formula + correct substitution.** |

Do not rearrange equations in algebraic form. Select the appropriate equation, substitute the given values then rearrange the equation to obtain the required unknown. This minimises the risk of a wrong substitution.

### *Use of the Data Sheet*

Clearly show where you have substituted a value from the data sheet. For example, **do not** leave $\mu_0$ in an equation. You must show the substitution $4\pi \times 10^{-7}$ in your equation.

When rounding, **do not** round the given values. For example, the mass of a proton = $1{\cdot}673 \times 10^{-27}$ kg **not** $1{\cdot}67 \times 10^{-27}$ kg.

### *Use of the Physics Relationship Sheet*

Although many of the required equations are given, it is better to know the basic equations to gain time in the examination.

### *"Show" questions*

Generally **all steps** for these must be given, even although they might seem obvious. **Do not assume that substitutions are obvious to the marker.**

All equations used must be stated separately and then clearly substituted if required. Many candidates will look at the end product and somehow end up with the required answer. The marker has to ensure that the path to the solution is clear. It is good practice to state why certain equations are used, explaining the Physics behind them.

### *Definitions*

Know and understand definitions given in the course. Definitions often come from the interpretation of an equation.

### *Diagrams*

Use a ruler and the appropriate labels. Angles will be important in certain diagrams. Too many candidates attempt to draw ray diagrams freehand.

### *Graphs*

Read the question and ensure you know what is being asked. Label your graph correctly and do not forget to label the origin.

### *"Explain" and "describe" questions*

These tend to be done poorly.

- Ensure all points are covered and read over again in order to check there are no mistakes. Try to be clear and to the point, highlighting the relevant Physics.
- Do not use up and down arrows in a description – this may help you in shorthand, but these must be translated to words.
- Be aware some answers require justification. No attempt at a justification can mean no marks awarded.

### *Two or more attempts at an answer*

Any attempt that you do not want the marker to consider should be scored out. Otherwise zero marks could be awarded.

Do not be tempted to give extra information that might be incorrect – all the marks may not be awarded if there are incorrect pieces of information. For example, this might include converting **incorrectly** m to nm in the last line of an answer, when it is not required.

**At the end of the exam, if you have time, quickly go over each answer and make sure you have the correct unit inserted.**

### *Experimental descriptions/planning*

You could well be called on to describe an experimental set up. Ensure your description is clear enough for another person to repeat it, and include a clearly labelled diagram.

### *Suggested improvements to the experimental procedure*

Look at the percentage uncertainties in the measured quantities and decide which is most significant. Suggest how the size of this uncertainty could be reduced – do not suggest using better apparatus! It might be better to repeat readings, so that random uncertainty is reduced or increase distances to reduce the percentage uncertainty in scale reading. There could be a systematic uncertainty that is affecting all readings. It really depends on the experiment. Use your judgement.

## Handling data

### *Relationships*

There are two methods to prove a direct or inverse relationship.

**Graphical approach**

Plot the graphs with the appropriate x and y values and look for a straight line – better plotted in pencil in case of mistakes. **Do not force a line through the origin!**

(A vs B for a direct relationship, C vs $\frac{1}{D}$ for an inverse relationship)

**Using the equation of a straight line,**
**$y = mx + c$.**

**Be aware that the gradient of the line can often lead to required values.**

Example: Finding the permeability of free space.

$$B = \frac{\mu_0 I}{2\pi r}$$

Express $B = \frac{\mu_0}{2\pi r} I$ in the form of $y = mx + c$ ($c = 0$)

By plotting the graph of B against I, the value of the gradient will give $\frac{\mu_0}{2\pi r}$

(Note: the line should not be forced through the origin.)

The value of C can often indicate a systematic uncertainty in the experiment. Ensure you are clear on how to calculate the gradient of a line.

**Algebraic approach**

If it appears that $A \propto B$ then calculate the value of $\frac{A}{B}$ **for all values**.

If these show that $\frac{A}{B} = k$ then the relationship holds.

If it appears that $C \propto \frac{1}{D}$ then calculate the value C.D **for all values**.

If these show that C.D = k then the relationship holds.

### *Uncertainties*

In this area, you must understand the following:

- systematic, calibration, scale reading (analogue and digital) and random uncertainties
- percentage/fractional uncertainties
- combinations – Pythagorean relationship
- absolute uncertainty in the final answer (give to one significant figure).

For your lab work, it is always useful and less time consuming to use a spreadsheet package to shortcut calculations, plot graphs and estimate uncertainties. Just ensure that if plotting a graph, it is at least half a page in size and use the smallest grid lines available.

### *Significant figures*

**Do not** round off in intermediate calculations, but round off in the final answer to an appropriate number of figures.

**Rounding off to three significant figures in the final answer will generally be acceptable.**

### *Prefixes*

Ensure you know all the required prefixes and be able to convert them to the correct power of 10.

### *Open-ended questions*

There will be two open-ended questions in the paper each worth 3 marks. Some candidates look upon these as mini essays.

Remember each is worth only 3 marks and they give the opportunity to demonstrate knowledge and understanding. **However, do not spend too long on these.** It might be better to revisit them at the end of the exam. Some students prefer to use bullet points to highlight the main areas of understanding.

Below is some advice on each unit. Obviously, all points in all units cannot be covered, but hopefully the following can give you a start in what to look out for.

## Unit 1 Rotational Motion and Astrophysics

### *Kinematic relationships*

#### Calculus — equations of motion

s = f(t) – given

Be clear that differentiating once gives the velocity, differentiating twice gives the acceleration.

a = f(t) – given

Integrating once gives the velocity, integrating twice gives the displacement. Remember to take into account the constant of integration each time by considering the limits or initial conditions.

#### Circular Motion

Make a two column table with the headings "Linear" and "Circular". On the left-hand side, write down all the equations you have come across in Higher, starting with speed and then the equations of motion. On the right-hand side, add all the equivalent equations for circular motion.

| Linear | Circular |
|---|---|
| $v = \frac{s}{t}$ | $\omega = \frac{\theta}{t}$ |
| $v = u + at$ | $\omega = \omega_0 + \alpha t$ |
| | |
| | |
| | |

Add to these as you progress through the course.

#### Central force

There will often be a question on **central** or **centripetal force**. Remember this is the force that acts on an object causing the object to follow a circular path.

**There is no outward (centrifugal) force acting on that object** and it would be good advice **not to mention this force** in your description.

It is worth noting that a fun fair ride might give the impression of feeling an outward force acting on a person, but this is an unreal force – they just think the force is outward. In fact, it is the inward force from the seat that enables you to follow the circular path.

The same will apply with a "loop the loop" ride where, at the top of the loop, the track and weight of the car supplies the required centripetal force.

**Moment of inertia** can be defined as a measure of the resistance to angular acceleration about a given axis (resistance to change).

#### Gravitation

#### Circular motion and planetary motion

The key to calculating the period, T, of motion of a planet or satellite lies with gravitational force supplying the central force, that is:

$\frac{GMm}{r^2} = mr\omega^2$ then use $\omega = \frac{2\pi}{T}$ to find $T$.

#### Escape velocity derivation

The starting point is the realisation when the object has escaped the pull of gravity **then $E_T = 0$.**

For example, $E_T = E_k + E_p = \frac{1}{2}mv^2 - \frac{GMm}{r} = 0$.

Rearrange the equation to obtain the **expression for v**.

(Note that $E_p = -\frac{GMm}{r}$)

#### General relativity

Remember the higher the altitude, then the gravitational field will be less, **which means clocks will run fast**. GPS systems have to take this into account, otherwise the position determined on Earth would be incorrect.

Mass curves spacetime which causes gravitational pull – rubber sheet analogy.

#### Stellar Physics

Become familiar with the Hertzsprung-Russell (H-R) diagram – stellar evolution.

## Unit 2 Quanta and Waves

### *Bohr model of the atom*

Many candidates omit the fact that the angular momentum of the electrons is quantised.

### *Wave particle duality*

Experimental evidence is often confused.

- Electron diffraction through crystals – particles as waves.
- Photoelectric effect – waves as particles.

## *Uncertainty principle*

It is difficult to measure something precisely without the measuring procedure affecting what you are trying to measure!

## *Simple harmonic motion: know the definition of SHM*

Be able to derive the velocity (differentiate once) and acceleration (differentiate again) given the expression for displacement:

$$y = A\sin\omega t \text{ or } x = A\cos\omega t$$

From v and a the expressions for the maximum velocity and maximum acceleration can be found. (These occur when the maximum value of cos or sin = 1.)

## *Waves*

**Wave Equation**

$$y = A\sin 2\pi\left(ft - \frac{x}{\lambda}\right)$$

Just knowing the coefficients of t and x in the wave equation allows the retrieval of frequency, wavelength and the speed of the wave,

$$\text{Coefficient of t} = 2\pi f$$

$$\text{Coefficient of x} = \frac{2\pi}{\lambda}$$

**Remember the $2\pi$!**

**Standing or Stationary waves**

Remember these are produced when a reflected wave interferes with the incident wave causing maxima (antinodes) and minima (nodes).

# Unit 3 Electromagnetism

## *Electric field strength*

**Parallel plates**

(uniform field between plates)

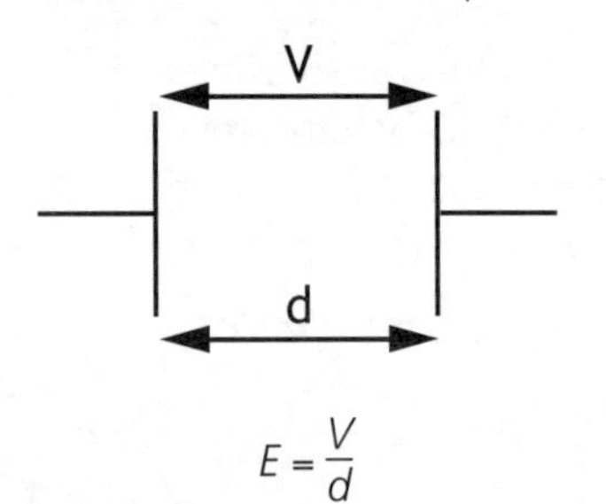


$$E = \frac{V}{d}$$

**Point charge**

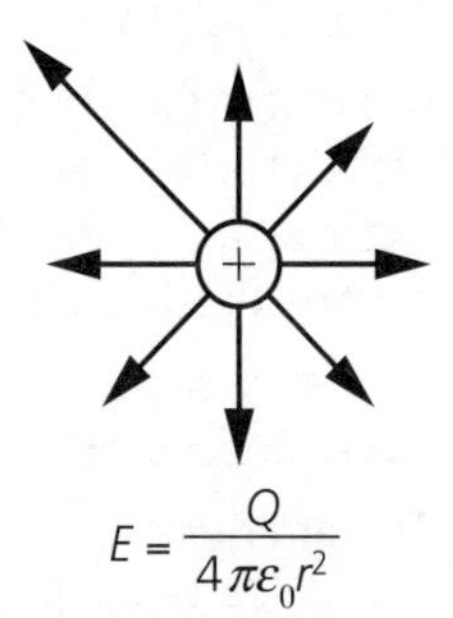


$$E = \frac{Q}{4\pi\varepsilon_0 r^2}$$

Candidates often mix up these formulae for electric field strength. They are quite different!

## *Ferromagnetism*

Certain materials can become magnetised due to the alignment of their magnetic dipoles.

## *Charged particles in a magnetic field*

A circular orbit will be produced if the charged particle cuts the field at 90°.

The central force is produced by the magnetic force acting on the particle.

$$\frac{mv^2}{r} = Bqv$$

If the charge enters the field at an angle **less than 90°** then the resultant path will be **helical**.

An example of this would be charged particles being deflected by the Earth's magnetic field to the North or South poles producing the borealis.

**Candidates should be able to explain why the resulting motion is helical.**

Treat this as two components of velocity:

$V_H$ = constant and moves in the direction of field lines – no force acting.

$V_V$ = 90° to the field which produces a central force causing the charge particle to follow a circular path.

## *Back emf in an inductor (coil)*

$$E = -L\frac{dI}{dt}$$

Remember that the back emf is produced by a changing current, which in turn produces a **changing magnetic field** throughout the coil. Many candidates omit this in their explanation.

Be aware that if asked to calculate the inductance L (or the rate of change of current) and a **negative answer** is obtained then **there has been an error. Invariably the fact that the induced emf is negative has not been taken into account**.

For example:

$$E = 9\text{V} \quad \frac{dI}{dt} = 5\,\text{As}^{-1}$$

$$E = -L\frac{dI}{dt}$$

$$9 = -L \times 5$$

$$L = -1{\cdot}8\text{ H}$$

**This is incorrect!**

$$E = -9\text{V} \quad \frac{dI}{dt} = 5\,\text{As}^{-1}$$

$$E = -L\frac{dI}{dt}$$

$$-9 = -L \times 5$$

$$L = 1{\cdot}8\text{ H}$$

**This is correct!**

ADVANCED HIGHER

# 2015 Specimen Question Paper

FOR OFFICIAL USE

AH

National Qualifications SPECIMEN ONLY

Mark

**SQ29/AH/01** **Physics**

Date — Not applicable

Duration — 2 hours 30 minutes

**Fill in these boxes and read what is printed below.**

Full name of centre

Town

Forename(s)

Surname

Number of seat

Date of birth

Day Month Year

DD MM YY

Scottish candidate number

**Total marks — 140**

Attempt ALL questions.

Reference may be made to the Physics Relationships Sheet and the Data Sheet on *Page two*.

Write your answers clearly in the spaces provided in this booklet. Additional space for answers is provided at the end of this booklet. If you use this space you must clearly identify the question number you are attempting. Any rough work must be written in this booklet. You should score through your rough work when you have written your final copy.

Use **blue** or **black** ink.

Before leaving the examination room you must give this booklet to the Invigilator; if you do not, you may lose all the marks for this paper.

**DATA SHEET**

COMMON PHYSICAL QUANTITIES

| *Quantity* | *Symbol* | *Value* | *Quantity* | *Symbol* | *Value* |
|---|---|---|---|---|---|
| Gravitational acceleration on Earth | $g$ | $9{\cdot}8\ \mathrm{m\,s^{-2}}$ | Mass of electron | $m_e$ | $9{\cdot}11 \times 10^{-31}$ kg |
| Radius of Earth | $R_E$ | $6{\cdot}4 \times 10^{6}$ m | Charge on electron | $e$ | $-1{\cdot}60 \times 10^{-19}$ C |
| Mass of Earth | $M_E$ | $6{\cdot}0 \times 10^{24}$ kg | Mass of neutron | $m_n$ | $1{\cdot}675 \times 10^{-27}$ kg |
| Mass of Moon | $M_M$ | $7{\cdot}3 \times 10^{22}$ kg | Mass of proton | $m_p$ | $1{\cdot}673 \times 10^{-27}$ kg |
| Radius of Moon | $R_M$ | $1{\cdot}7 \times 10^{6}$ m | Mass of alpha particle | $m_a$ | $6{\cdot}645 \times 10^{-27}$ kg |
| Mean Radius of Moon Orbit | | $3{\cdot}84 \times 10^{8}$ m | Charge on alpha particle | | $3{\cdot}20 \times 10^{-19}$ C |
| Solar radius | | $6{\cdot}955 \times 10^{8}$ m | Planck's constant | $h$ | $6{\cdot}63 \times 10^{-34}$ J s |
| Mass of Sun | | $2{\cdot}0 \times 10^{30}$ kg | Permittivity of free space | $\varepsilon_0$ | $8{\cdot}85 \times 10^{-12}\ \mathrm{F\,m^{-1}}$ |
| 1 AU | | $1{\cdot}5 \times 10^{11}$ m | Permeability of free space | $\mu_0$ | $4\pi \times 10^{-7}\ \mathrm{H\,m^{-1}}$ |
| Stefan-Boltzmann constant | $\sigma$ | $5{\cdot}67 \times 10^{-8}\ \mathrm{W\ m^{-2}\,K^{-4}}$ | Speed of light in vacuum | $c$ | $3{\cdot}0 \times 10^{8}\ \mathrm{m\,s^{-1}}$ |
| Universal constant of gravitation | $G$ | $6{\cdot}67 \times 10^{-11}\ \mathrm{m^3\,kg^{-1}\,s^{-2}}$ | Speed of sound in air | $v$ | $3{\cdot}4 \times 10^{2}\ \mathrm{m\,s^{-1}}$ |

REFRACTIVE INDICES

The refractive indices refer to sodium light of wavelength 589 nm and to substances at a temperature of 273 K.

| *Substance* | *Refractive index* | *Substance* | *Refractive index* |
|---|---|---|---|
| Diamond | 2·42 | Glycerol | 1·47 |
| Glass | 1·51 | Water | 1·33 |
| Ice | 1·31 | Air | 1·00 |
| Perspex | 1·49 | Magnesium Fluoride | 1·38 |

SPECTRAL LINES

| *Element* | *Wavelength*/nm | *Colour* |
|---|---|---|
| Hydrogen | 656 | Red |
| | 486 | Blue-green |
| | 434 | Blue-violet |
| | 410 | Violet |
| | 397 | Ultraviolet |
| | 389 | Ultraviolet |
| Sodium | 589 | Yellow |

| *Element* | *Wavelength*/nm | *Colour* |
|---|---|---|
| Cadmium | 644 | Red |
| | 509 | Green |
| | 480 | Blue |

*Lasers*

| *Element* | *Wavelength*/nm | *Colour* |
|---|---|---|
| Carbon dioxide | 9550 } 10590 | Infrared |
| Helium-neon | 633 | Red |

PROPERTIES OF SELECTED MATERIALS

| *Substance* | *Density*/ kg m$^{-3}$ | *Melting Point*/ K | *Boiling Point*/K | *Specific Heat Capacity*/ J kg$^{-1}$ K$^{-1}$ | *Specific Latent Heat of Fusion*/ J kg$^{-1}$ | *Specific Latent Heat of Vaporisation*/ J kg$^{-1}$ |
|---|---|---|---|---|---|---|
| Aluminium | $2{\cdot}70 \times 10^{3}$ | 933 | 2623 | $9{\cdot}02 \times 10^{2}$ | $3{\cdot}95 \times 10^{5}$ | .... |
| Copper | $8{\cdot}96 \times 10^{3}$ | 1357 | 2853 | $3{\cdot}86 \times 10^{2}$ | $2{\cdot}05 \times 10^{5}$ | .... |
| Glass | $2{\cdot}60 \times 10^{3}$ | 1400 | .... | $6{\cdot}70 \times 10^{2}$ | .... | .... |
| Ice | $9{\cdot}20 \times 10^{2}$ | 273 | .... | $2{\cdot}10 \times 10^{3}$ | $3{\cdot}34 \times 10^{5}$ | .... |
| Glycerol | $1{\cdot}26 \times 10^{3}$ | 291 | 563 | $2{\cdot}43 \times 10^{3}$ | $1{\cdot}81 \times 10^{5}$ | $8{\cdot}30 \times 10^{5}$ |
| Methanol | $7{\cdot}91 \times 10^{2}$ | 175 | 338 | $2{\cdot}52 \times 10^{3}$ | $9{\cdot}9 \times 10^{4}$ | $1{\cdot}12 \times 10^{6}$ |
| Sea Water | $1{\cdot}02 \times 10^{3}$ | 264 | 377 | $3{\cdot}93 \times 10^{3}$ | .... | .... |
| Water | $1{\cdot}00 \times 10^{3}$ | 273 | 373 | $4{\cdot}19 \times 10^{3}$ | $3{\cdot}34 \times 10^{5}$ | $2{\cdot}26 \times 10^{6}$ |
| Air | 1·29 | .... | .... | .... | .... | .... |
| Hydrogen | $9{\cdot}0 \times 10^{-2}$ | 14 | 20 | $1{\cdot}43 \times 10^{4}$ | .... | $4{\cdot}50 \times 10^{5}$ |
| Nitrogen | 1·25 | 63 | 77 | $1{\cdot}04 \times 10^{3}$ | .... | $2{\cdot}00 \times 10^{5}$ |
| Oxygen | 1·43 | 55 | 90 | $9{\cdot}18 \times 10^{2}$ | .... | $2{\cdot}40 \times 10^{4}$ |

The gas densities refer to a temperature of 273 K and a pressure of $1{\cdot}01 \times 10^{5}$ Pa.

**Total marks — 140 marks**

**Attempt ALL questions**

MARKS | DO NOT WRITE IN THIS MARGIN

1. Water is removed from clothes during the spin cycle of a washing machine. The drum holding the clothes has a maximum spin rate of 1250 revolutions per minute.

Figure 1A

(a) Show that the maximum angular velocity of the drum is 131 rad $s^{-1}$. **2**

*Space for working and answer*

(b) At the start of a spin cycle the drum has an angular velocity of 7·50 rad $s^{-1}$. It then takes 12·0 seconds to accelerate to the maximum angular velocity.

(i) Calculate the angular acceleration of the drum during the 12·0 seconds, assuming the acceleration is uniform. **3**

*Space for working and answer*

MARKS | DO NOT WRITE IN THIS MARGIN

**1. (b) (continued)**

(ii) Determine how many revolutions the drum will make during the 12·0 seconds. **5**

*Space for working and answer*

(c) When the drum is rotating at maximum angular velocity, an item of wet clothing of mass $1{\cdot}5 \times 10^{-2}$ kg rotates at a distance of 0·28 m from the axis of rotation as shown in Figure 1B.

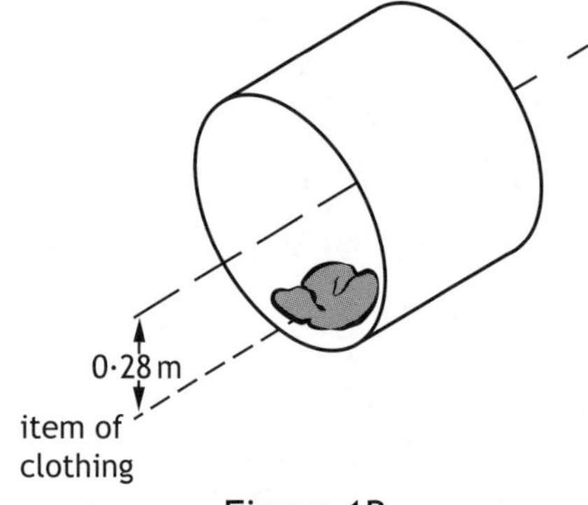


Figure 1B

Calculate the centripetal force acting on the item of clothing. **3**

*Space for working and answer*

MARKS | DO NOT WRITE IN THIS MARGIN

**1. (continued)**

(d) The outer surface of the drum has small holes as shown in Figure 1C. These holes allow most of the water to be removed.

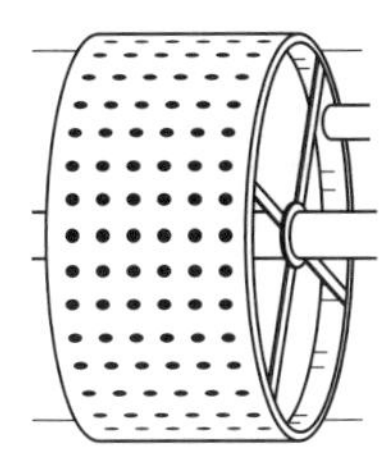

Figure 1C

(i) Explain why the water separates from the item of clothing during the spin cycle. **2**

(ii) The drum rotates in an anticlockwise direction. Indicate on Figure 1D the direction taken by a water droplet as it leaves the drum. **1**

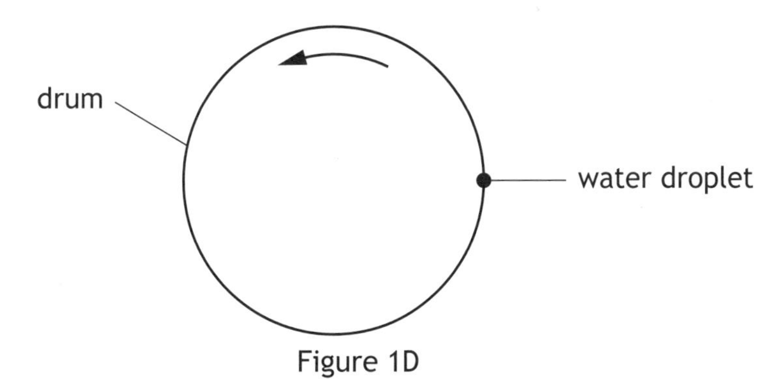


Figure 1D

(iii) Explain what happens to the value of the force on an item of clothing inside the drum as it rotates at its maximum angular velocity. **2**

MARKS | DO NOT WRITE IN THIS MARGIN

**2.** A disc of mass 6·0 kg and radius 0·50 m is allowed to rotate freely about its central axis as shown in Figure 2A.

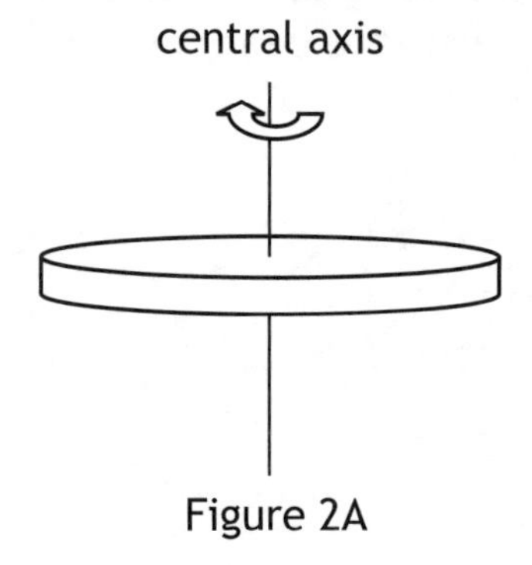


Figure 2A

(a) Show that the moment of inertia of the disc is 0·75 kg $m^2$. **2**

*Space for working and answer*

(b) The disc is rotating with an angular velocity of 12 rad $s^{-1}$. A cube of mass 2·0 kg is then dropped onto the disc. The cube remains at a distance of 0·40 m from the axis of rotation as shown in Figure 2B.

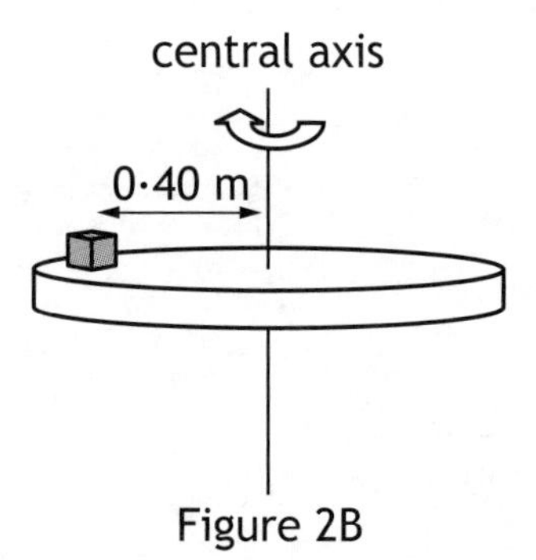


Figure 2B

MARKS | DO NOT WRITE IN THIS MARGIN

**2. (b) (continued)**

(i) Determine the total moment of inertia of the disc and cube. **3**

*Space for working and answer*

(ii) Calculate the angular velocity of the disc after the cube lands. **3**

*Space for working and answer*

(iii) State **one** assumption you have made in your response to b(ii). **1**

(c) The cube is removed and the disc is again made to rotate with a constant angular velocity of 12 rad $s^{-1}$. A sphere of mass 2·0 kg is then dropped onto the disc at a distance of 0·40 m from the axis as shown in Figure 2C.

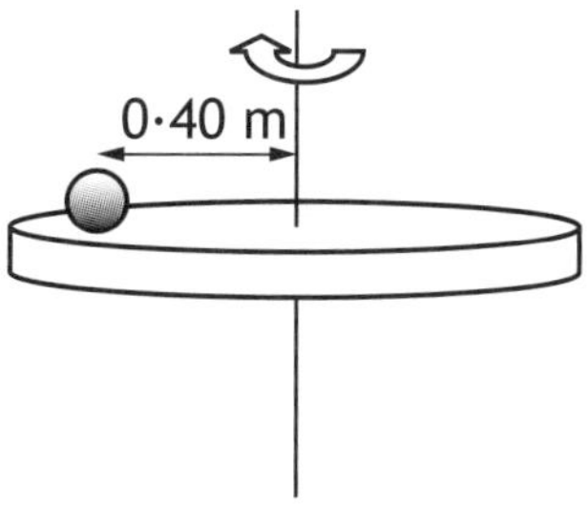

Figure 2C

State whether the resulting angular velocity of the disc is greater than, the same as, or less than, the value calculated in b(ii).

You must justify your answer. **2**

MARKS DO NOT WRITE IN THIS MARGIN

**3.** The International Space Station (ISS) is in orbit around the Earth.

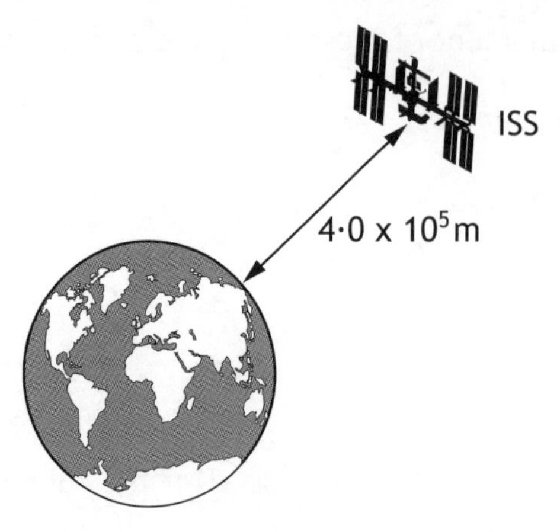


Figure 3A

(a) (i) The gravitational pull of the Earth keeps the ISS in orbit.

Show that for an orbit of radius $r$ the period $T$ is given by the expression

$$T = 2\pi\sqrt{\frac{r^3}{GM_E}}$$

where the symbols have their usual meaning. **2**

(ii) Calculate the period of orbit of the ISS when it is at an altitude of $4{\cdot}0 \times 10^5$ m above the surface of the Earth. **2**

*Space for working and answer*

MARKS DO NOT WRITE IN THIS MARGIN

3. **(continued)**

(b) The graph in Figure 3B shows how the altitude of the ISS has varied over time. Reductions in altitude are due to the drag of the Earth's atmosphere acting on the ISS.

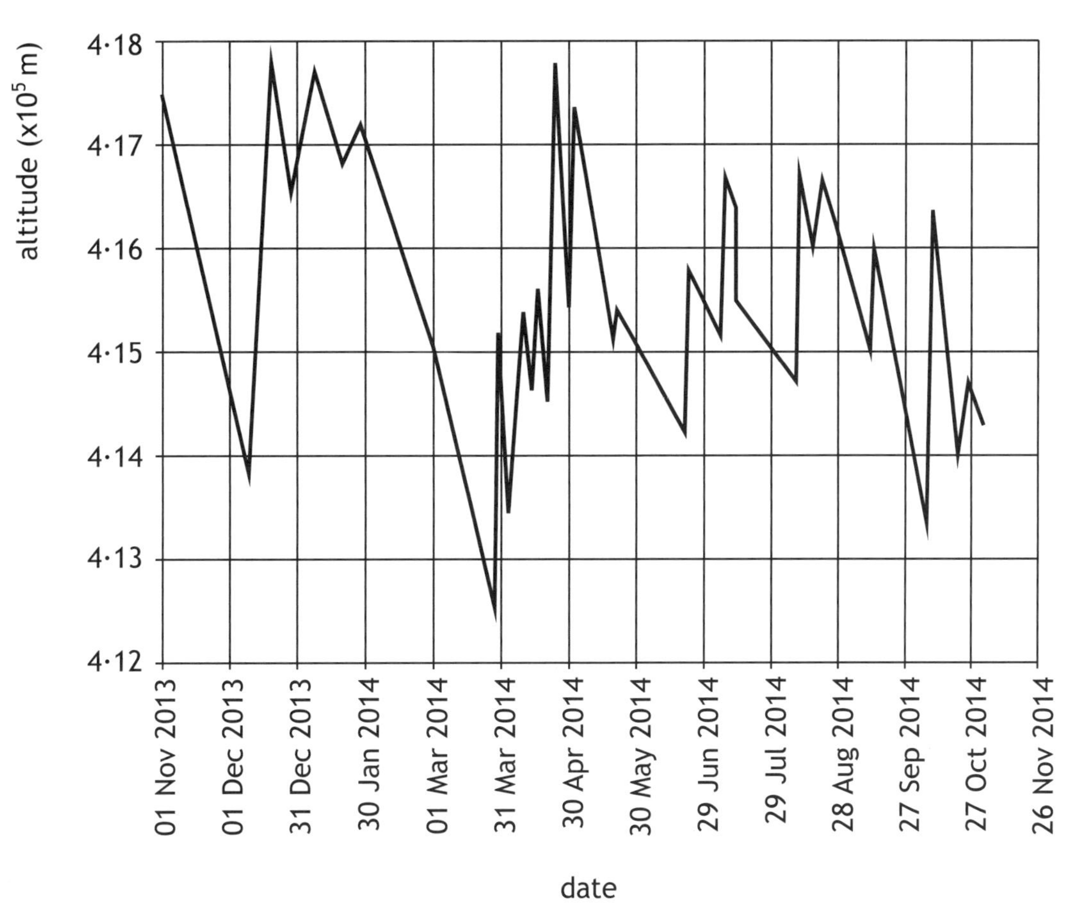


Figure 3B

(i) Determine the value of Earth's gravitational field strength at the ISS on 1 March 2014. **4**

*Space for working and answer*

MARKS DO NOT WRITE IN THIS MARGIN

3. (b) (continued)

(ii) In 2011 the average altitude of the ISS was increased from 350 km to 400 km.

Give an advantage of operating the ISS at this higher altitude. 1

(c) Clocks designed to operate on the ISS are synchronised with clocks on Earth before they go into space. On the ISS a correction factor is necessary for the clocks to remain synchronised with clocks on the Earth.

Explain why this correction factor is necessary. 2

MARKS DO NOT WRITE IN THIS MARGIN

4. The constellation Orion, shown in Figure 4A, is a common sight in the winter sky above Scotland.

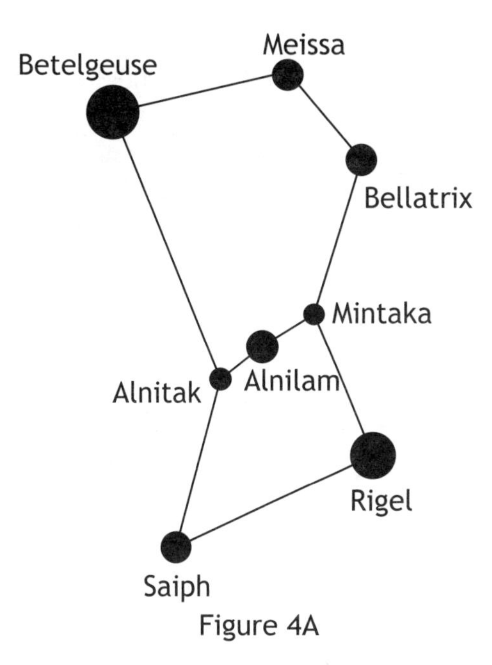


Figure 4A

Two of the stars in this constellation are known as Betelgeuse and Rigel. Their positions are shown on the Hertzsprung–Russell (H–R) diagram in Figure 4B.

MARKS | DO NOT WRITE IN THIS MARGIN

**4. (continued)**

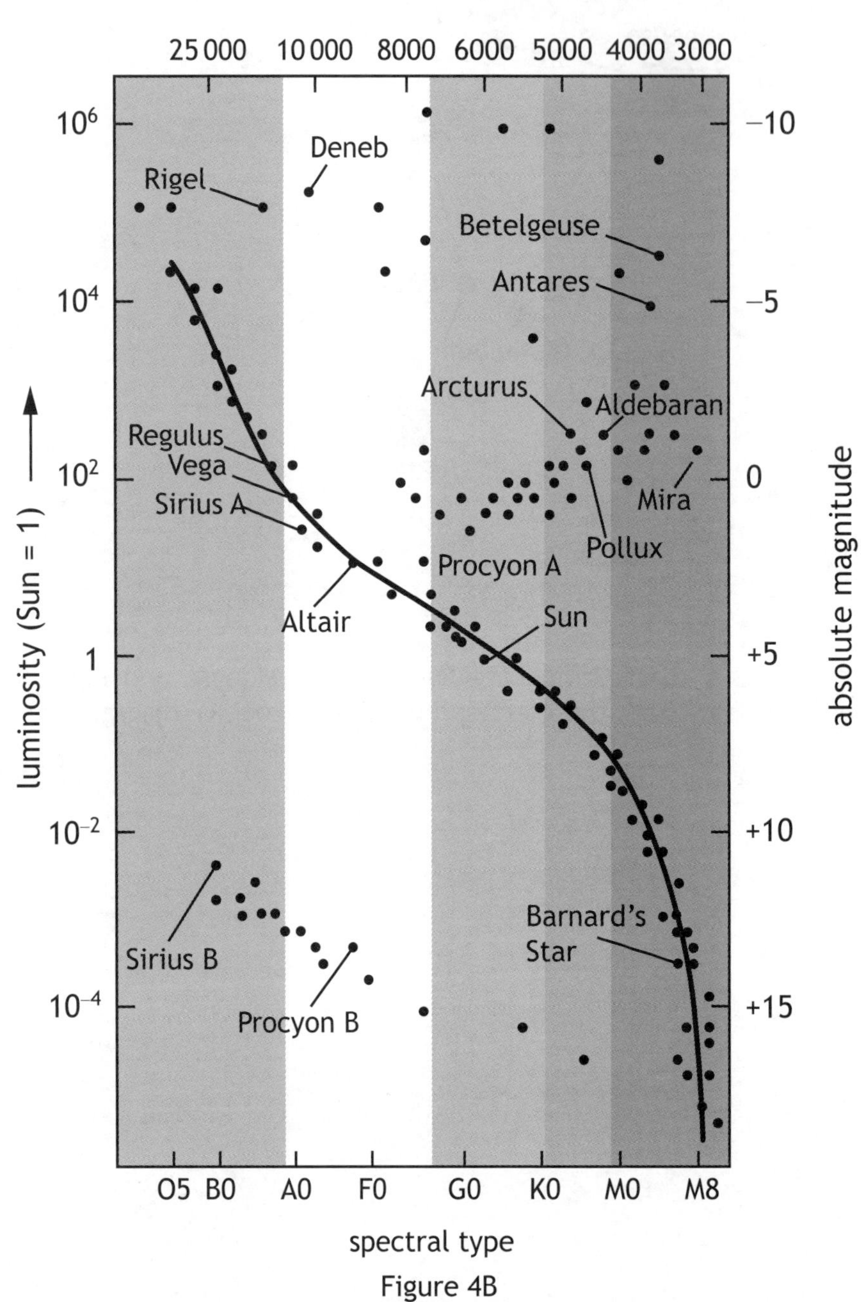


Figure 4B

(a) Using the H—R diagram, predict the colour of Betelgeuse. **1**

MARKS DO NOT WRITE IN THIS MARGIN

**4. (continued)**

(b) The table shows some of the physical properties of Rigel.

| *Property of Rigel* | |
|---|---|
| Surface temperature | $(1 \cdot 20 \pm 0 \cdot 05) \times 10^4$ K |
| Radius | $(5 \cdot 49 \pm 0 \cdot 50) \times 10^{10}$ m |
| Mass | $18 \pm 1$ solar masses |
| Distance to Earth | $773 \pm 150$ light years |

(i) (A) Calculate the luminosity of Rigel. **3**

*Space for working and answer*

(B) State the assumption made in your calculation. **1**

(ii) Calculate the absolute uncertainty in the value of the luminosity of Rigel. **4**

*Space for working and answer*

MARKS | DO NOT WRITE IN THIS MARGIN

**4. (continued)**

(c) Calculate the apparent brightness of Rigel as observed from the Earth. **4**

*Space for working and answer*

(d) Betelgeuse is not on the Main Sequence region of the H—R diagram. Describe the changes that have taken place in Betelgeuse since leaving the Main Sequence. **2**

MARKS DO NOT WRITE IN THIS MARGIN

**5.** Figure 5A shows a snowboarder in a half pipe. The snowboarder is moving between positions P and Q. The total mass of the snowboarder and board is 85 kg.

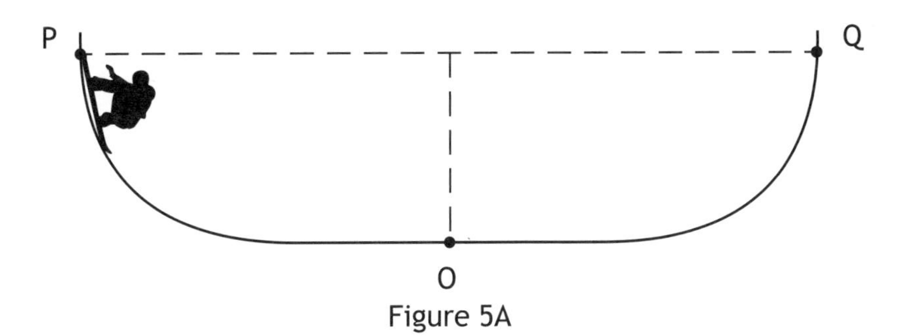

Figure 5A

A student attempts to model the motion of the snowboarder as simple harmonic motion (SHM).

The student uses measurements of amplitude and period to produce the displacement-time graph shown in Figure 5B.

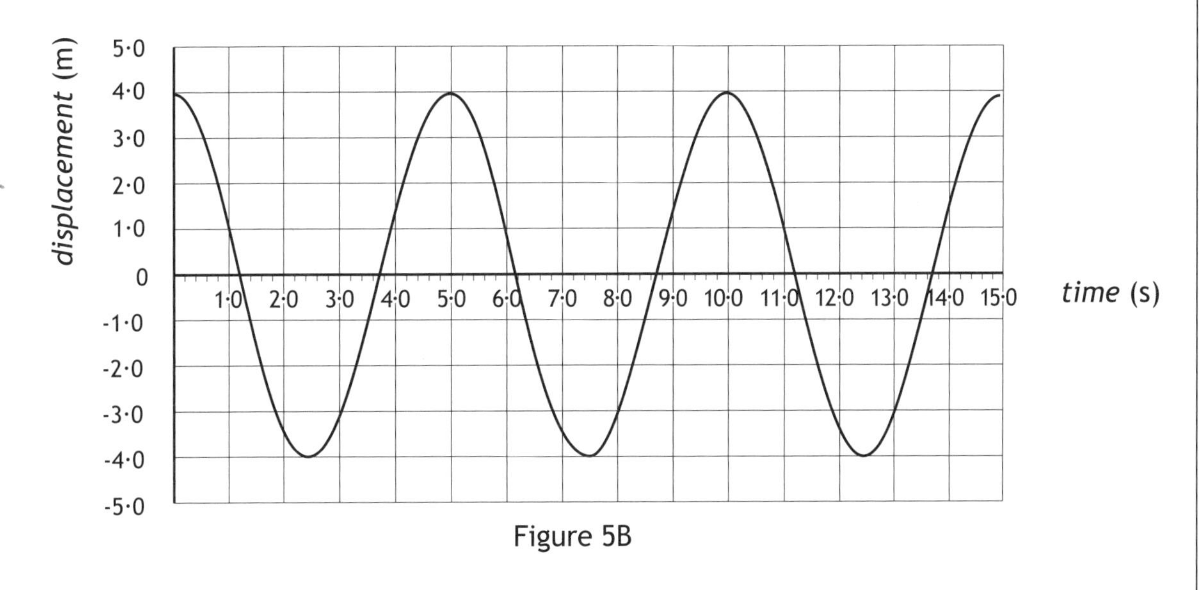

Figure 5B

(a) (i) State what is meant by the term *simple harmonic motion*. **1**

MARKS | DO NOT WRITE IN THIS MARGIN

**5. (a) (continued)**

(ii) Determine the angular frequency of the motion. **4**

*Space for working and answer*

(iii) Calculate the maximum acceleration experienced by the snowboarder on the halfpipe. **3**

*Space for working and answer*

(iv) Sketch a velocity-time graph for one period of this motion.

Numerical values are required on both axes. **3**

You may wish to use the square-ruled paper on *Page thirty*.

(v) Calculate the maximum potential energy of the snowboarder. **3**

*Space for working and answer*

MARKS DO NOT WRITE IN THIS MARGIN

**5. (continued)**

(b) Detailed video analysis shows that the snowboarder's motion is not fully described by the SHM model.

Using your knowledge of physics, comment on possible reasons for this discrepancy. **3**

MARKS | DO NOT WRITE IN THIS MARGIN

6. The Bohr model of the hydrogen atom consists of a single electron orbiting a single proton. Due to the quantisation of angular momentum, in this model, the electron can only orbit at particular radii.

Figure 6A shows an electron with principal quantum number $n = 1$.

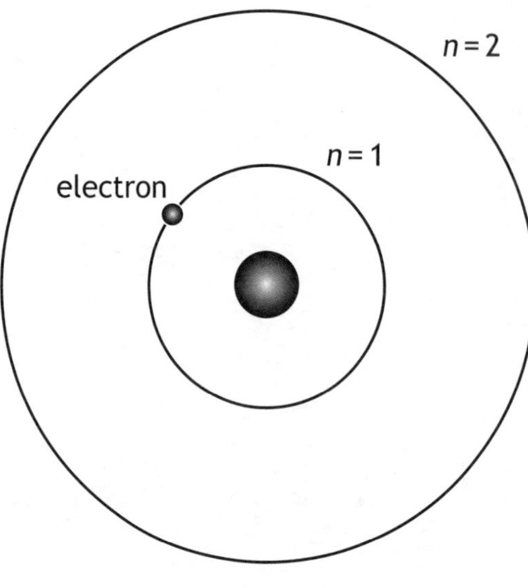


Figure 6A

(a) Explain what gives rise to the centripetal force acting on the electron. **1**

MARKS DO NOT WRITE IN THIS MARGIN

**6. (continued)**

(b) (i) Show that the kinetic energy of the electron is given by

$$E_k = \frac{e^2}{8\pi\varepsilon_0 r}$$

where the symbols have their usual meaning. **2**

(ii) Calculate the kinetic energy for an electron with orbital radius 0·21 nm. **2**

*Space for working and answer*

(c) Calculate the principal quantum number for an electron with angular momentum $4{\cdot}22 \times 10^{-34}\,\text{kg m}^2\text{s}^{-1}$. **3**

*Space for working and answer*

MARKS | DO NOT WRITE IN THIS MARGIN

**6. (continued)**

(d) Heisenberg's uncertainty principle addresses some of the limitations of classical physics in describing quantum phenomena.

(i) The uncertainty in an experimental measurement of the momentum of an electron in a hydrogen atom was determined to be $\pm 1{\cdot}5 \times 10^{-26}$ kg m s$^{-1}$ .

Calculate the minimum uncertainty in the position of the electron. **3**

*Space for working and answer*

(ii) In a scanning tunnelling microscope (STM) a sharp metallic tip is brought very close to the surface of a conductor. As the tip is moved back and forth, an electric current can be detected due to the movement ("tunnelling") of electrons across the air gap between the tip and the conductor, as shown in Figure 6B.

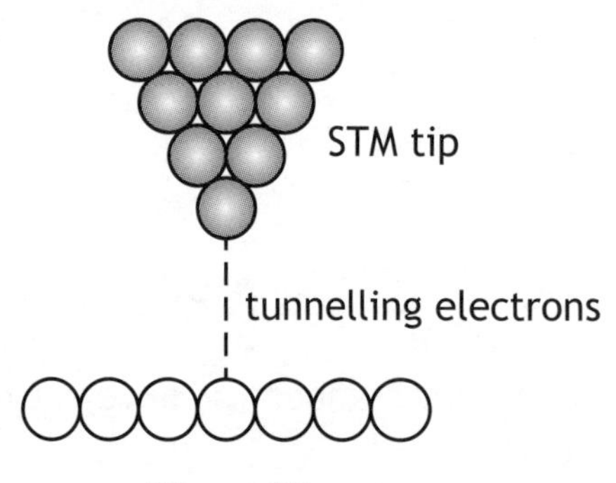


Figure 6B

According to classical physics, electrons should not be able to cross the gap as the kinetic energy of each electron is insufficient to overcome the repulsion between electrons in the STM tip and the surface.

Explain why an electron is able to cross the gap. **3**

MARKS | DO NOT WRITE IN THIS MARGIN

7. When a microwave oven is switched on a stationary wave is formed inside the oven.

(a) Explain how a stationary wave is formed. **1**

(b) A student carries out an experiment to determine the speed of light using a microwave oven. The turntable is removed from the oven and bread covered in butter is placed inside. The oven is switched on for a short time, after which the student observes that the butter has melted only in certain spots, as shown in Figure 7A.

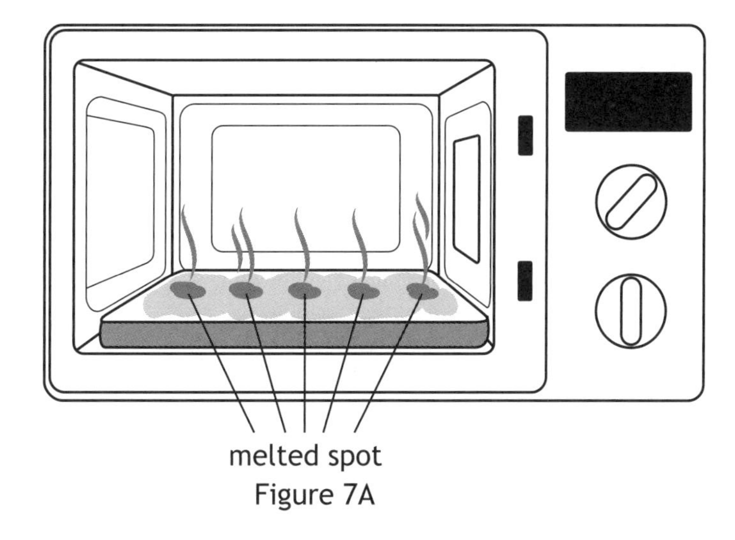


Figure 7A

Explain why the butter has melted in certain spots and not in others. **2**

MARKS DO NOT WRITE IN THIS MARGIN

**7. (continued)**

(c) The student measures the distance between the first hot spot and fifth hot spot as 264 mm.

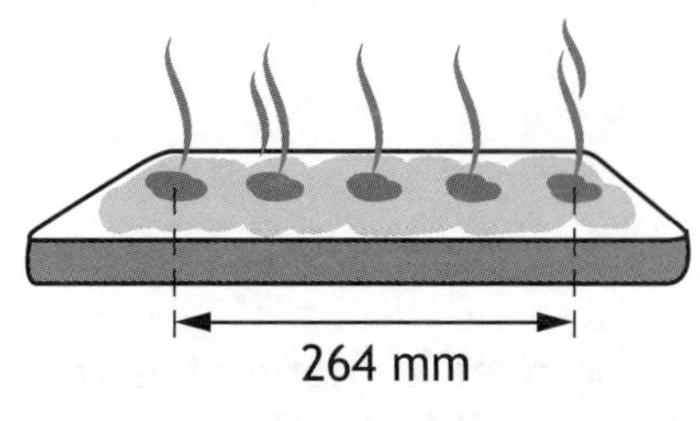


Figure 7B

From the data obtained by the student determine the wavelength of the microwaves. **2**

*Space for working and answer*

(d) The quoted value for the frequency of the microwaves is 2·45 GHz. The student calculates the speed of light using data from the experiment.

Show that the value obtained by the student for the speed of light is $3{\cdot}23 \times 10^8\,m\,s^{-1}$. **2**

*Space for working and answer*

(e) The student repeats the experiment and obtains the following values for the speed of light,

$3{\cdot}26 \times 10^8\,m\,s^{-1}$, $3{\cdot}19 \times 10^8\,m\,s^{-1}$, $3{\cdot}23 \times 10^8\,m\,s^{-1}$, $3{\cdot}21 \times 10^8\,m\,s^{-1}$.

Comment on both the accuracy and precision of the student's results. **2**

MARKS DO NOT WRITE IN THIS MARGIN

**8.** A beam of electrons is incident on a grating as shown in Figure 8A.

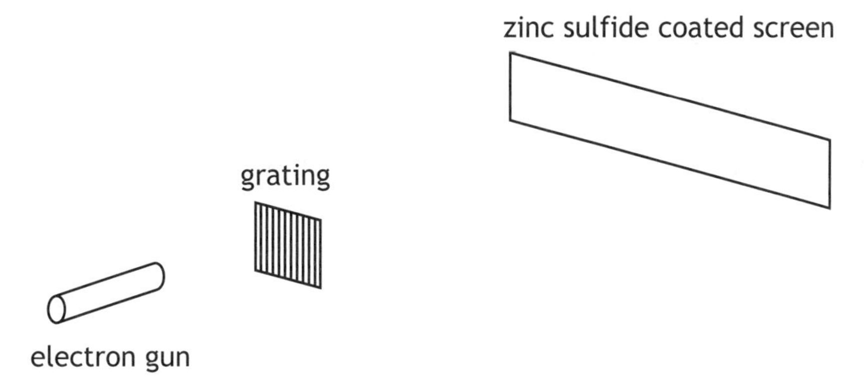


Figure 8A

(a) After passing through the grating the electrons are incident on a zinc sulfide coated screen. The coating emits light when struck by electrons.

Describe the pattern observed on the screen. **1**

(b) Scientists perform similar experiments with large molecules. One such molecule is buckminsterfullerene (C60) with a mass of $1\cdot20 \times 10^{-24}$ kg.

For C60 molecules with a velocity of $220\,\text{m s}^{-1}$ estimate the slit spacing required to produce a pattern comparable to that observed for the electrons. You must justify your answer by calculation. **4**

*Space for working and answer*

MARKS DO NOT WRITE IN THIS MARGIN

9. As part of a physics project a student carried out experiments to obtain values for the permeability of free space and the permittivity of free space.

The results obtained by the student were

permeability of free space, $\mu_0 = (1 \cdot 32 \pm 0 \cdot 05) \times 10^{-6}\,\text{H m}^{-1}$

permittivity of free space, $\varepsilon_0 = (8 \cdot 93 \pm 0 \cdot 07) \times 10^{-12}\,\text{F m}^{-1}$

(a) State the number of significant figures in the value of each result. **1**

(b) Use these results to determine a value for the speed of light.

Your answer must be consistent with (a). **3**

*Space for working and answer*

(c) (i) Determine which of the uncertainties obtained by the student is more significant for the calculation of the speed of light.

You must justify your answer by calculation. **3**

*Space for working and answer*

(ii) Calculate the absolute uncertainty in the value obtained for the speed of light. **2**

*Space for working and answer*

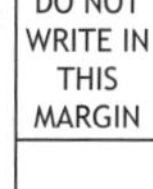


MARKS

**10.** (a) Two point charges $Q_1$ and $Q_2$ are separated by a distance of 0·60 m as shown in Figure 10A. The charge on $Q_1$ is −8·0 nC. The electric field strength at point X is zero.

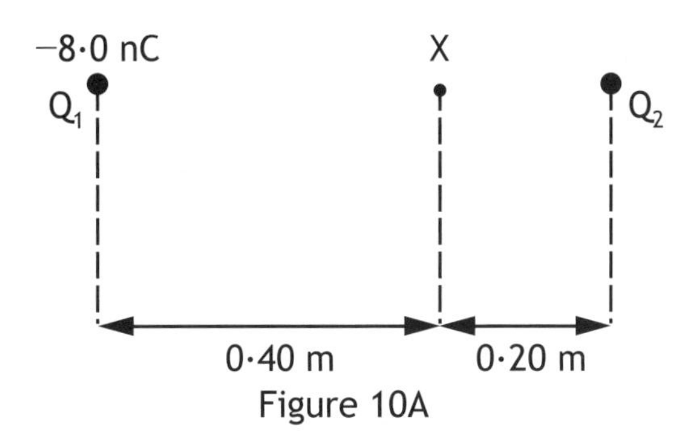


Figure 10A

(i) State what is meant by *electric field strength*. **1**

(ii) Show that the charge on $Q_2$ is −2·0 nC. **2**

*Space for working and answer*

(iii) Calculate the electrical potential at point X. **5**

*Space for working and answer*

MARKS DO NOT WRITE IN THIS MARGIN

**10. (continued)**

(b)

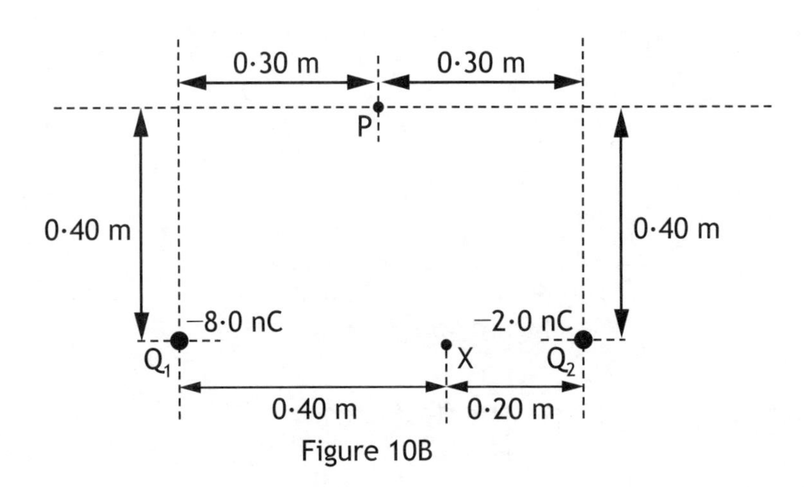

Figure 10B

(i) Calculate the electrical potential at point P. **3**

*Space for working and answer*

(ii) Determine the energy required to move a charge of +1·0 nC from point X to point P. **4**

*Space for working and answer*

MARKS DO NOT WRITE IN THIS MARGIN

**11.** The Nobel prize winning physicist Richard Feynman once stated "things on a small scale behave nothing like things on a large scale".

Using your knowledge of physics, comment on his statement. **3**

MARKS 

**12.** A student carries out a series of experiments to investigate properties of capacitors in a.c. circuits.

(a) The student connects a 5·0 μF capacitor to an a.c. supply of e.m.f. 15 $V_{rms}$ and negligible internal resistance as shown in Figure 12A.

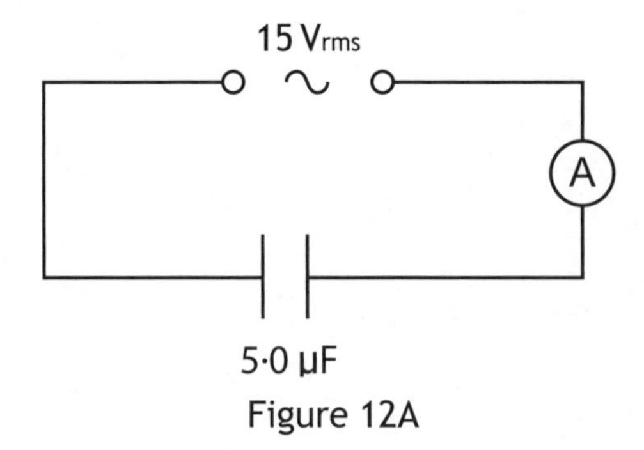



Figure 12A

The frequency of the a.c. supply is 65 Hz.

(i) Calculate the reactance of the capacitor. **3**

*Space for working and answer*

(ii) Determine the value of the current in the circuit. **3**

*Space for working and answer*

MARKS | DO NOT WRITE IN THIS MARGIN

**12. (continued)**

(b) The student uses the following circuit to determine the capacitance of a second capacitor.

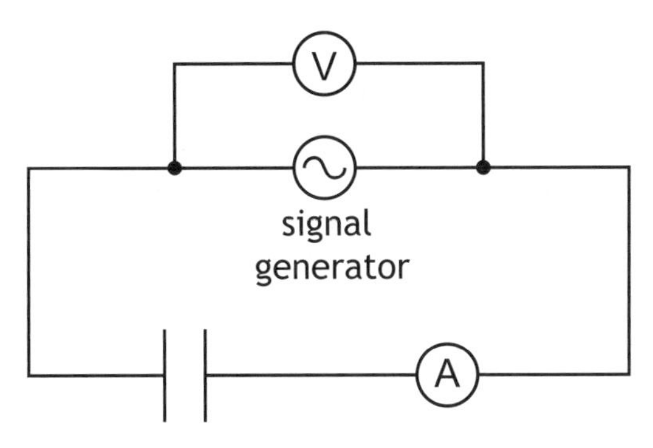


Figure 12B

The student obtains the following data.

| *Reactance* (Ω) | *Frequency* (Hz) |
|---|---|
| $1{\cdot}60 \times 10^6$ | 10 |
| $6{\cdot}47 \times 10^5$ | 40 |
| $2{\cdot}99 \times 10^5$ | 100 |
| $1{\cdot}52 \times 10^5$ | 200 |
| $6{\cdot}35 \times 10^4$ | 500 |
| $3{\cdot}18 \times 10^4$ | 1000 |

(i) On the square-ruled paper on *Page thirty*, plot a graph that would be suitable to determine the capacitance. **4**

(ii) Use your graph to determine the capacitance of this capacitor. **3**

*Space for working and answer*

**[END OF SPECIMEN QUESTION PAPER]**

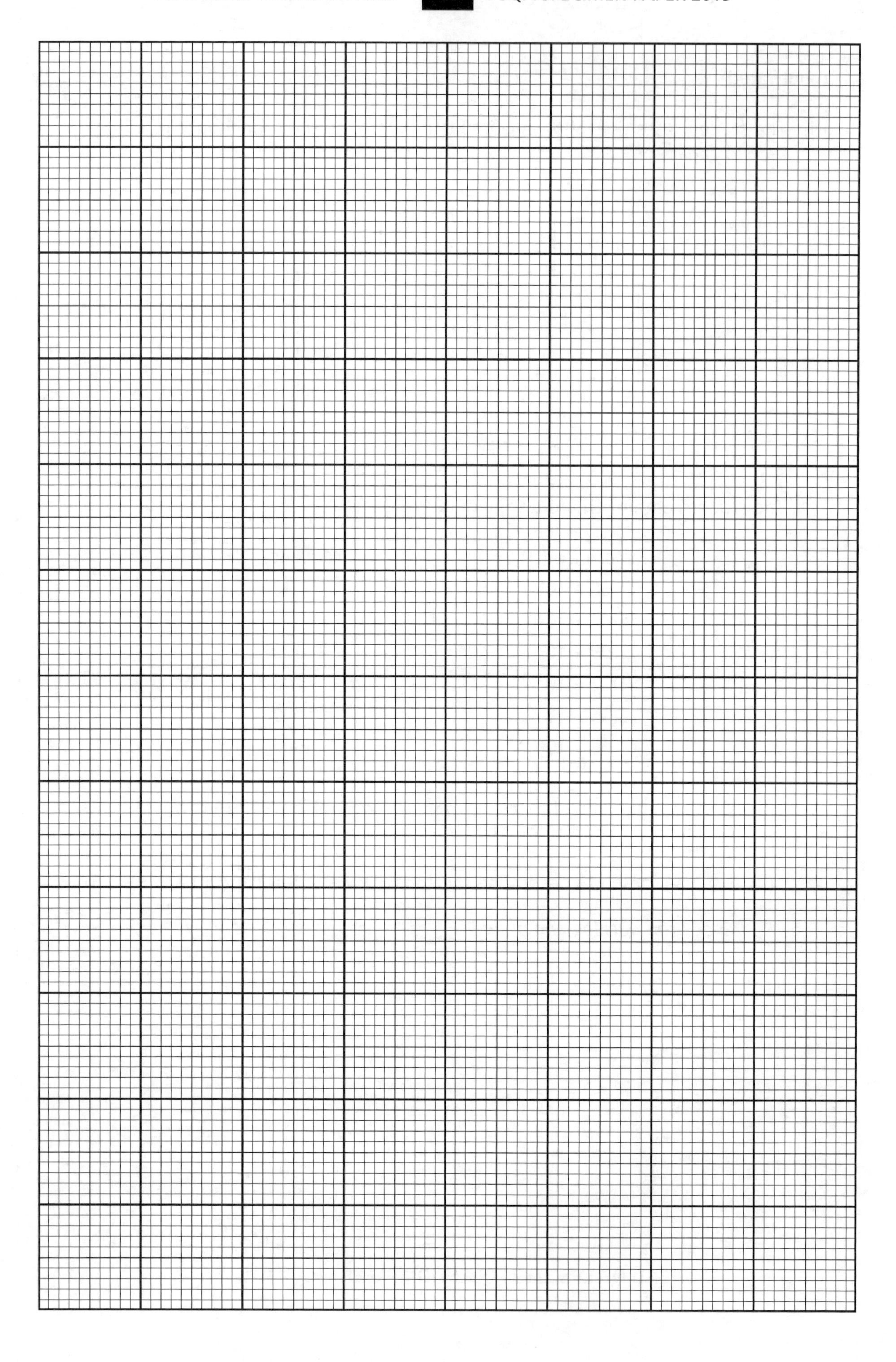

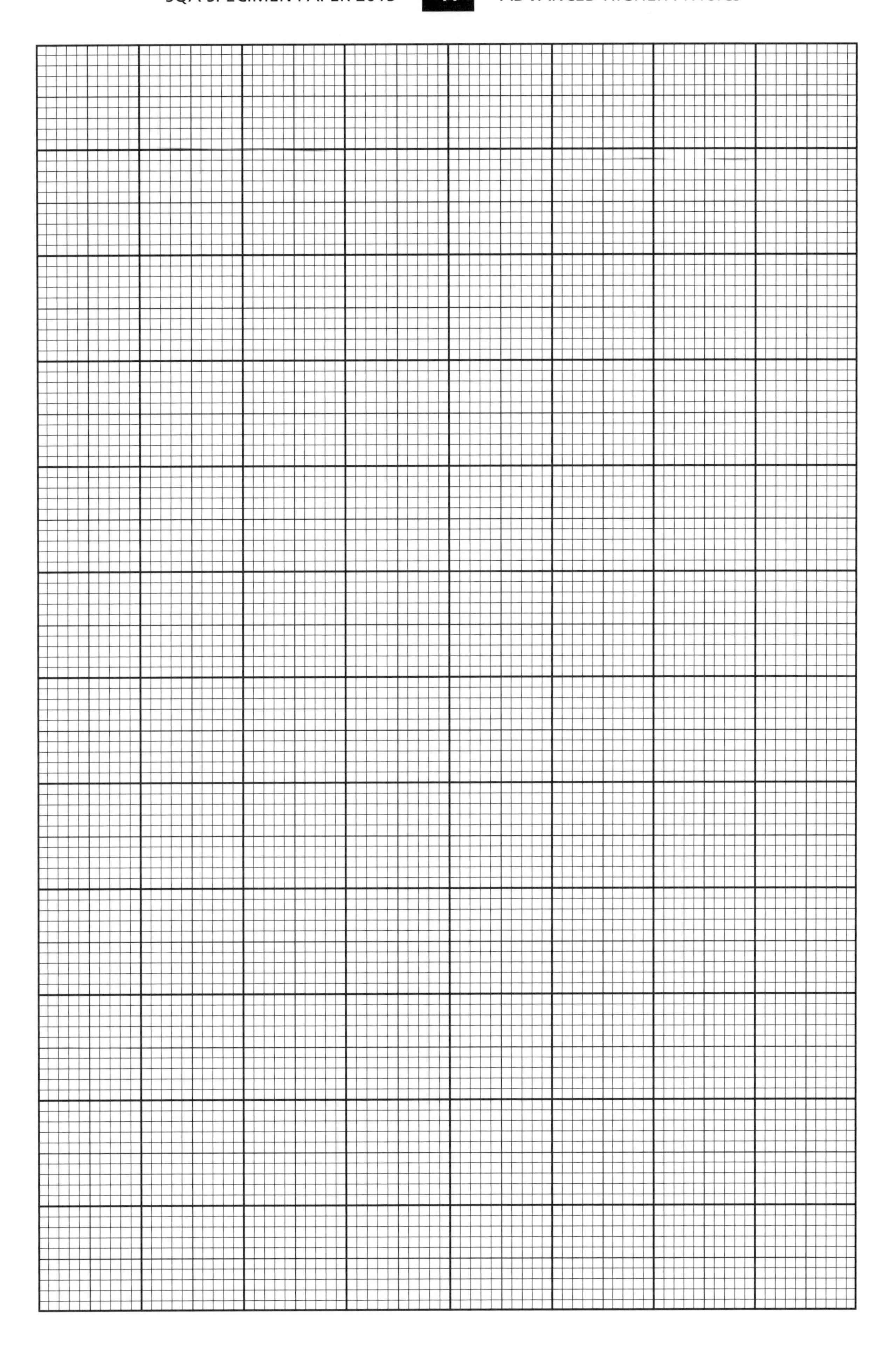

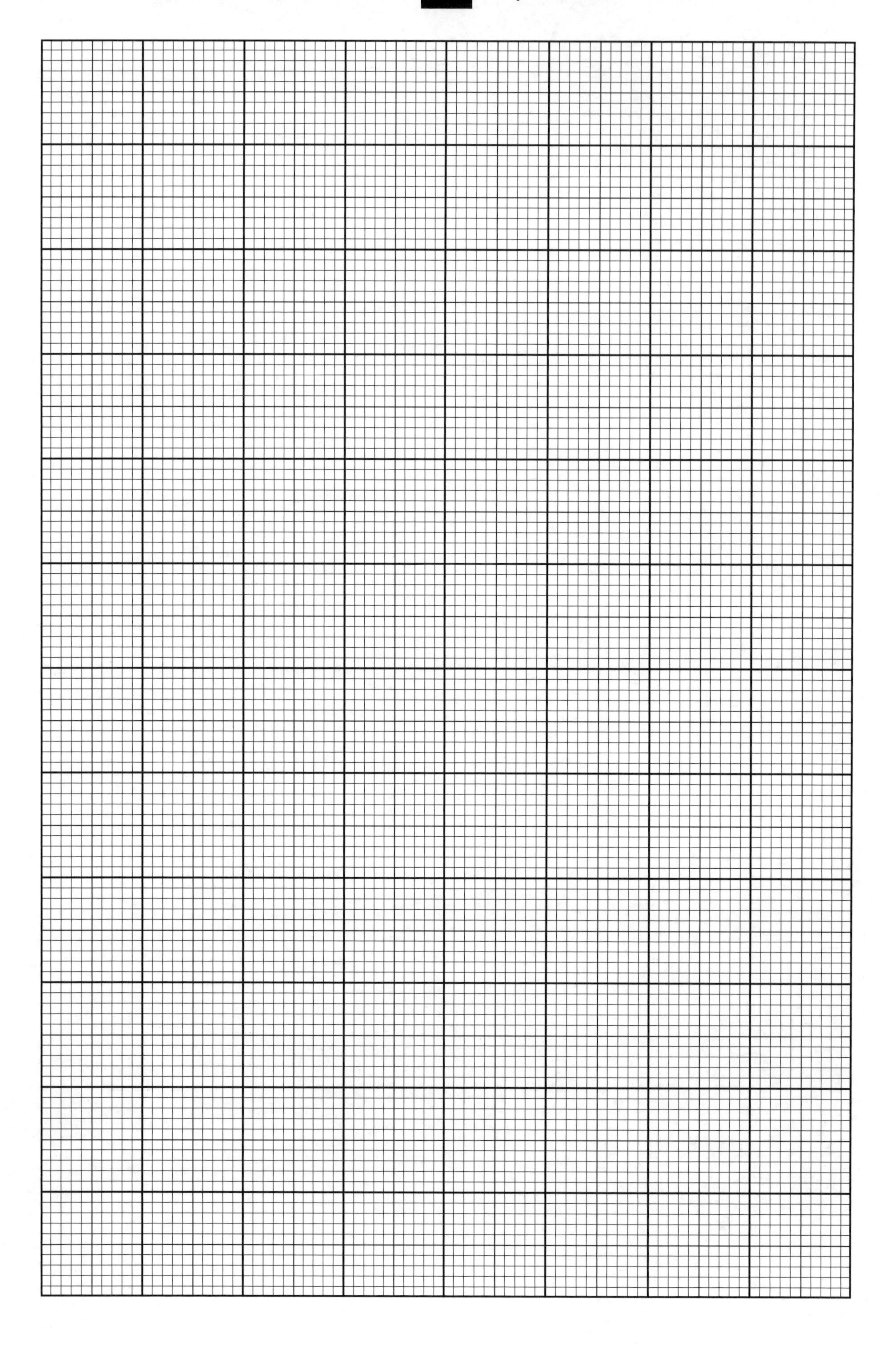

MARKS | DO NOT WRITE IN THIS MARGIN

**ADDITIONAL SPACE FOR ANSWERS AND ROUGH WORK**

**ADDITIONAL SPACE FOR ANSWERS AND ROUGH WORK**

## ADVANCED HIGHER

# Model Paper 1

Whilst this Model Paper has been specially commissioned by Hodder Gibson for use as practice for the Advanced Higher (for Curriculum for Excellence) exams, the key reference document remains the SQA Specimen Paper 2015 and the SQA Exam Paper 2016.

FOR OFFICIAL USE

National
Qualifications
MODEL PAPER 1

Mark

# Physics

Duration — 2 hours 30 minutes

**Fill in these boxes and read what is printed below.**

Full name of centre

Town

Forename(s)

Surname

Number of seat

Date of birth

Day Month Year

D D M M Y Y

Scottish candidate number

**Total marks — 140**

Attempt ALL questions.

Reference may be made to the Physics Relationships Sheet and the Data Sheet on *Page two*.

Write your answers clearly in the spaces provided in this booklet. Additional space for answers is provided at the end of this booklet. If you use this space you must clearly identify the question number you are attempting. Any rough work must be written in this booklet. You should score through your rough work when you have written your final copy.

Use **blue** or **black** ink.

Before leaving the examination room you must give this booklet to the Invigilator; if you do not, you may lose all the marks for this paper.

**DATA SHEET**

COMMON PHYSICAL QUANTITIES

| Quantity | Symbol | Value | Quantity | Symbol | Value |
|---|---|---|---|---|---|
| Gravitational acceleration on Earth | $g$ | $9{\cdot}8\ \text{m s}^{-2}$ | Mass of electron | $m_e$ | $9{\cdot}11 \times 10^{-31}$ kg |
| Radius of Earth | $R_E$ | $6{\cdot}4 \times 10^{6}$ m | Charge on electron | $e$ | $-1{\cdot}60 \times 10^{-19}$ C |
| Mass of Earth | $M_E$ | $6{\cdot}0 \times 10^{24}$ kg | Mass of neutron | $m_n$ | $1{\cdot}675 \times 10^{-27}$ kg |
| Mass of Moon | $M_M$ | $7{\cdot}3 \times 10^{22}$ kg | Mass of proton | $m_p$ | $1{\cdot}673 \times 10^{-27}$ kg |
| Radius of Moon | $R_M$ | $1{\cdot}7 \times 10^{6}$ m | Mass of alpha particle | $m_a$ | $6{\cdot}645 \times 10^{-27}$ kg |
| Mean Radius of Moon Orbit | | $3{\cdot}84 \times 10^{8}$ m | Charge on alpha particle | | $3{\cdot}20 \times 10^{-19}$ C |
| Solar radius | | $6{\cdot}955 \times 10^{8}$ m | Planck's constant | $h$ | $6{\cdot}63 \times 10^{-34}$ J s |
| Mass of Sun | | $2{\cdot}0 \times 10^{30}$ kg | Permittivity of free space | $\varepsilon_0$ | $8{\cdot}85 \times 10^{-12}\ \text{F m}^{-1}$ |
| 1 AU | | $1{\cdot}5 \times 10^{11}$ m | Permeability of free space | $\mu_0$ | $4\pi \times 10^{-7}\ \text{H m}^{-1}$ |
| Stefan-Boltzmann constant | $\sigma$ | $5{\cdot}67 \times 10^{-8}\ \text{W m}^{-2}\text{K}^{-4}$ | Speed of light in vacuum | $c$ | $3{\cdot}0 \times 10^{8}\ \text{m s}^{-1}$ |
| Universal constant of gravitation | $G$ | $6{\cdot}67 \times 10^{-11}\ \text{m}^3\ \text{kg}^{-1}\text{s}^{-2}$ | Speed of sound in air | $v$ | $3{\cdot}4 \times 10^{2}\ \text{m s}^{-1}$ |

REFRACTIVE INDICES

The refractive indices refer to sodium light of wavelength 589 nm and to substances at a temperature of 273 K.

| Substance | Refractive index | Substance | Refractive index |
|---|---|---|---|
| Diamond | 2·42 | Glycerol | 1·47 |
| Glass | 1·51 | Water | 1·33 |
| Ice | 1·31 | Air | 1·00 |
| Perspex | 1·49 | Magnesium Fluoride | 1·38 |

SPECTRAL LINES

| Element | Wavelength/nm | Colour | Element | Wavelength/nm | Colour |
|---|---|---|---|---|---|
| Hydrogen | 656 | Red | Cadmium | 644 | Red |
| | 486 | Blue-green | | 509 | Green |
| | 434 | Blue-violet | | 480 | Blue |
| | 410 | Violet | *Lasers* | | |
| | 397 | Ultraviolet | *Element* | *Wavelength/nm* | *Colour* |
| | 389 | Ultraviolet | Carbon dioxide | 9550 } 10590 } | Infrared |
| Sodium | 589 | Yellow | Helium-neon | 633 | Red |

PROPERTIES OF SELECTED MATERIALS

| Substance | Density/ kg m$^{-3}$ | Melting Point/ K | Boiling Point/K | Specific Heat Capacity/ J kg$^{-1}$ K$^{-1}$ | Specific Latent Heat of Fusion/ J kg$^{-1}$ | Specific Latent Heat of Vaporisation/ J kg$^{-1}$ |
|---|---|---|---|---|---|---|
| Aluminium | $2{\cdot}70 \times 10^{3}$ | 933 | 2623 | $9{\cdot}02 \times 10^{2}$ | $3{\cdot}95 \times 10^{5}$ | . . . . |
| Copper | $8{\cdot}96 \times 10^{3}$ | 1357 | 2853 | $3{\cdot}86 \times 10^{2}$ | $2{\cdot}05 \times 10^{5}$ | . . . . |
| Glass | $2{\cdot}60 \times 10^{3}$ | 1400 | . . . . | $6{\cdot}70 \times 10^{2}$ | . . . . | . . . . |
| Ice | $9{\cdot}20 \times 10^{2}$ | 273 | . . . . | $2{\cdot}10 \times 10^{3}$ | $3{\cdot}34 \times 10^{5}$ | . . . . |
| Glycerol | $1{\cdot}26 \times 10^{3}$ | 291 | 563 | $2{\cdot}43 \times 10^{3}$ | $1{\cdot}81 \times 10^{5}$ | $8{\cdot}30 \times 10^{5}$ |
| Methanol | $7{\cdot}91 \times 10^{2}$ | 175 | 338 | $2{\cdot}52 \times 10^{3}$ | $9{\cdot}9 \times 10^{4}$ | $1{\cdot}12 \times 10^{6}$ |
| Sea Water | $1{\cdot}02 \times 10^{3}$ | 264 | 377 | $3{\cdot}93 \times 10^{3}$ | . . . . | . . . . |
| Water | $1{\cdot}00 \times 10^{3}$ | 273 | 373 | $4{\cdot}19 \times 10^{3}$ | $3{\cdot}34 \times 10^{5}$ | $2{\cdot}26 \times 10^{6}$ |
| Air | 1·29 | . . . . | . . . . | . . . . | . . . . | . . . . |
| Hydrogen | $9{\cdot}0 \times 10^{-2}$ | 14 | 20 | $1{\cdot}43 \times 10^{4}$ | . . . . | $4{\cdot}50 \times 10^{5}$ |
| Nitrogen | 1·25 | 63 | 77 | $1{\cdot}04 \times 10^{3}$ | . . . . | $2{\cdot}00 \times 10^{5}$ |
| Oxygen | 1·43 | 55 | 90 | $9{\cdot}18 \times 10^{2}$ | . . . . | $2{\cdot}40 \times 10^{4}$ |

The gas densities refer to a temperature of 273 K and a pressure of $1{\cdot}01 \times 10^{5}$ Pa.

**Total marks — 140 marks**

**Attempt ALL questions**

MARKS | DO NOT WRITE IN THIS MARGIN

**1.** (a) A student investigates the average acceleration of a radio-controlled car.

She marks the points *A* and *B* on a straight track, as shown in Figure 1A and measures the distance *AB* using a measuring tape.

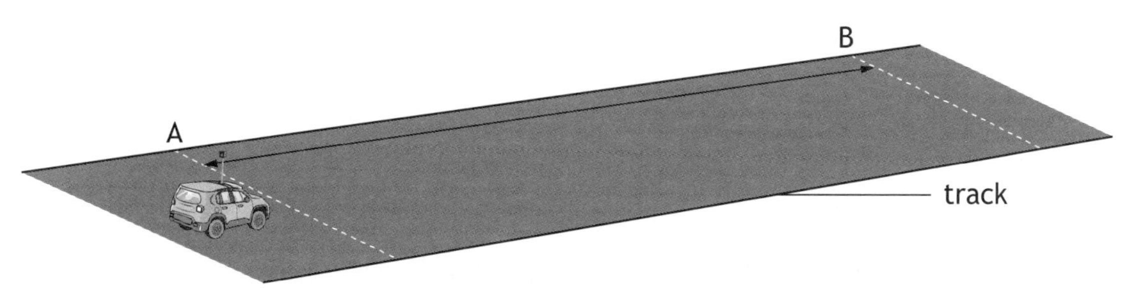


Figure 1A

She places the car at *A* and uses the radio control to accelerate the car.

The car starts from rest and accelerates in a straight line along the track to *B*.

Using a stopwatch, the student measures the time for the car to travel the distance *AB*.

She repeats this several times and obtains the following results.

Distance *AB* = (3·54 ± 0·01) m.

Stopwatch readings: 2·53 s; 2·29 s; 2·34 s; 2·36 s; 2·65 s; 2·53 s.

(i) Starting with the appropriate equation of motion, show that the acceleration of the car is given by

$$a = \frac{2s}{t^2}$$

where the symbols have their usual meanings. **1**

(ii) Calculate the mean value of the car's acceleration. **2**

MARKS DO NOT WRITE IN THIS MARGIN

**1. (a) (continued)**

(iii) Calculate the random uncertainty in the time measurement. **2**

(iv) Calculate the percentage uncertainty in the mean acceleration. **3**

(v) Express the numerical result of her investigation in the form final value ± absolute uncertainty. **2**

(b) The student now places the car, which has a mass of 2·5 kg, on a horizontal circular track as shown in Figure 1B.

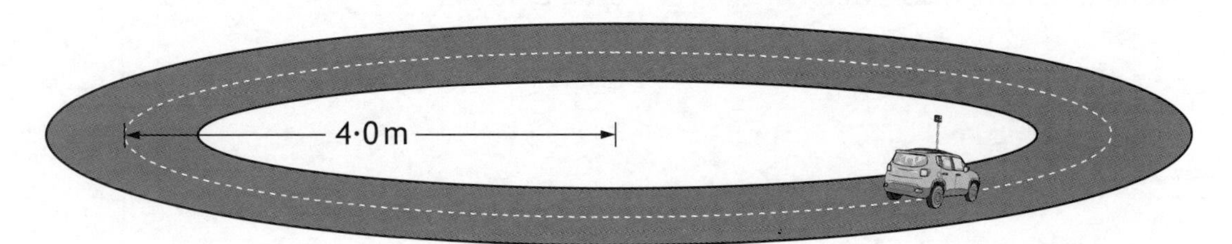


Figure 1B

She uses the radio control to make the car travel with a constant speed of 6·0 $m\,s^{-1}$ around a circular path of radius 4·0 m.

MARKS DO NOT WRITE IN THIS MARGIN

**1. (b) (continued)**

(i) Calculate the car's radial acceleration. **3**

(ii) The radial friction between the car's tyres and the track has a maximum value of 23 N.

Show that this force is sufficient to prevent the car skidding off the track. **3**

(c) The student now places the car on a banked track as shown in Figure 1C.

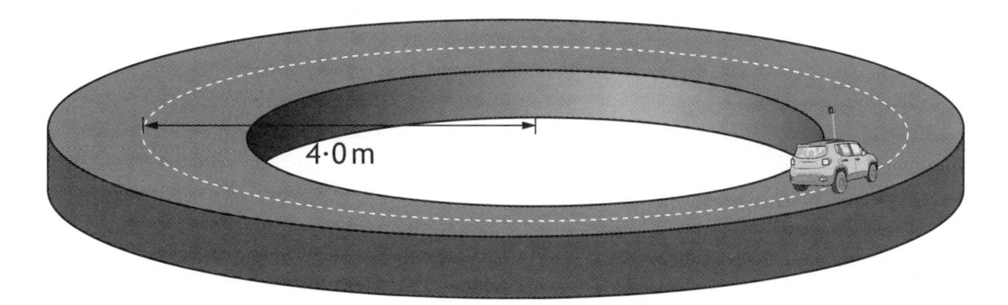


Figure 1C

She again uses the radio control to make the car follow a circular path of radius 4·0 m.

Explain why the car can now travel much faster than 6·0 m $s^{-1}$ without skidding off the track. **2**

MARKS DO NOT WRITE IN THIS MARGIN

2. (a) A mass $m$ rotates at a distance $r$ from a fixed axis.

State the expression for its moment of inertia. **1**

(b) A toy gyroscope consists of an axle, a narrow ring and spokes as shown in Figure 2A.

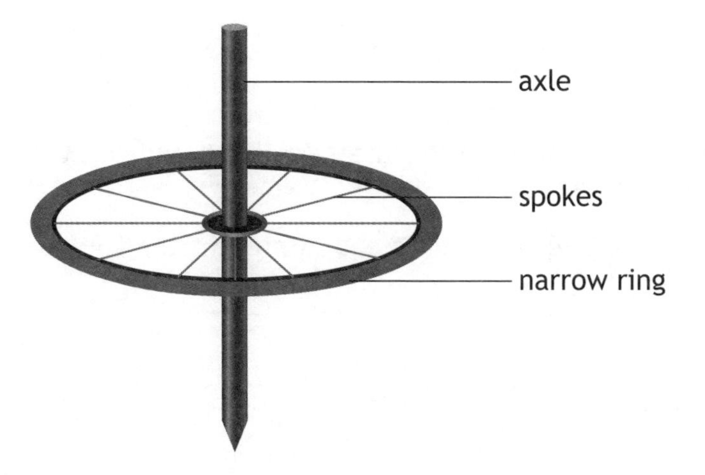


Figure 2A

The mass of the axle and spokes is negligible compared to the mass of the narrow ring.

The narrow ring has a mass of 1·5 kg and an average radius of 0·20 m.

Show that the moment of inertia of the gyroscope is 0·060 kg m$^2$. **1**

(c) The axle of the gyroscope has a radius of 4·0 mm.

The gyroscope is made to spin using a thin cord. A 0·50 m length of thin cord is wound round the axle, as shown in Figure 2B.

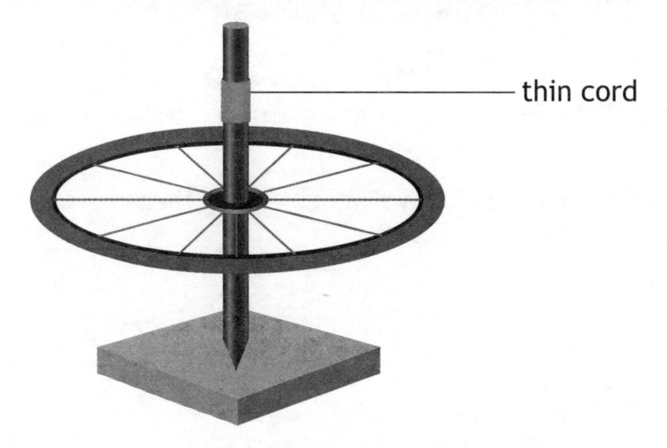


Figure 2B

The cord is pulled with a steady horizontal force of 25 N.

A constant frictional torque of 0·070 Nm opposes the motion of the gyroscope.

MARKS | DO NOT WRITE IN THIS MARGIN

**2. (c) (continued)**

(i) Calculate the resultant torque acting on the gyroscope. **3**

(ii) Calculate the angular acceleration of the gyroscope. **3**

(iii) Show that the angular displacement of the gyroscope is 125 radians just as the cord fully unwinds. **3**

(iv) Calculate the maximum angular velocity of the gyroscope. **3**

(v) After the cord has fully unwound, the frictional torque remains constant. How long does it take for the angular velocity of the gyroscope to decrease to $4{\cdot}2\,\text{rad}\,\text{s}^{-1}$? **4**

MARKS DO NOT WRITE IN THIS MARGIN

3. The Moon orbits the Earth due to the gravitational force between them.

(a) Show that the magnitude of the gravitational force between the Earth and the Moon is $2 \cdot 0 \times 10^{20}$ N. **2**

(b) Hence calculate the tangential speed of the Moon in its orbit around the Earth. **3**

(c) Calculate the potential energy of the Moon in its orbit. **3**

(d) Hence calculate the total energy of the Moon in its orbit. **3**

MARKS DO NOT WRITE IN THIS MARGIN

4. (a) With reference to General Relativity, explain why the Moon orbits the Earth. **2**

(b) General Relativity also predicts gravitational lensing.

Figure 4 shows the relative positions of Earth, a massive object and a distant star (not to scale).

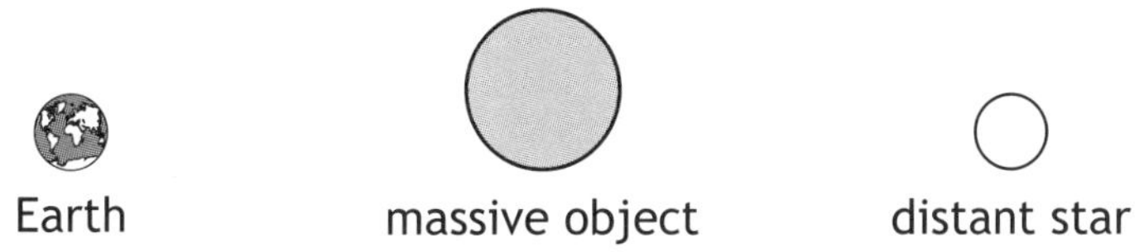

Earth massive object distant star

Figure 4

On the diagram show:

(i) the path of light from the star to Earth **1**

(ii) the observed position of the star from Earth. **1**

(c) Two students visit the tallest building on Earth. Student *A* takes a lift to the top of the building while student *B* waits at the bottom. General Relativity predicts that time will not pass at the same rate for both students.

For which student does time pass at a slower rate?

You must justify your answer. **2**

MARKS DO NOT WRITE IN THIS MARGIN

5. A funfair attendant, Jim, steps off a roundabout ride as it spins around.

Tom, seated on the roundabout, states

*"I saw Jim flying outwards from the ride so there must be an outward (centrifugal) force acting on him."*

Use your knowledge of Physics to comment on the Tom's statement. **3**

MARKS DO NOT WRITE IN THIS MARGIN

**6.** (a) (i) Electrons exhibit wave-like behaviour.

Give one example of experimental evidence which supports this statement. **1**

(ii) Electrons can also exhibit particle-like behaviour.

Give one example of experimental evidence which supports this statement. **1**

(b) De Broglie showed that it is possible to calculate a wavelength for a moving object.

A tennis ball of mass 60 g is served at 55 m $s^{-1}$.

(i) Calculate the de Broglie wavelength for this ball. **3**

(ii) Explain why wave-like properties are not observed for this ball. **1**

MARKS DO NOT WRITE IN THIS MARGIN

**7.** One of the key ideas in Quantum Theory is the Heisenberg Uncertainty Principle.

The uncertainty in the position of a particle can be estimated as its de Broglie wavelength.

(a) An electron has an average speed of $3 \cdot 2 \times 10^6$ m s$^{-1}$.

(i) Calculate the minimum uncertainty in the momentum of this electron. **4**

(ii) It is not possible to measure accurately the position of an electron using visible light. Describe the effect of using a beam of X-rays rather than visible light on the measurement of the electron's position and momentum. Justify your answer. **3**

(b) Polonium 212 decays by alpha emission. The energy required for an alpha particle to escape from the Polonium nucleus is 26 MeV. Prior to emission, alpha particles in the nucleus have an energy of 8·78 MeV.

With reference to the Uncertainty Principle, explain how this process can occur. **2**

MARKS | DO NOT WRITE IN THIS MARGIN

**8.** A student uses the simple pendulum to find gravitational field strength, $g$.

She measures the time for 10 complete swings of the pendulum for different lengths, $l$.

This is repeated 5 times for each length.

The relationship between the period, $T$, of the pendulum and the length is given by

$$T = 2\pi\sqrt{\frac{l}{g}}$$

The pupil plots the graph shown below.

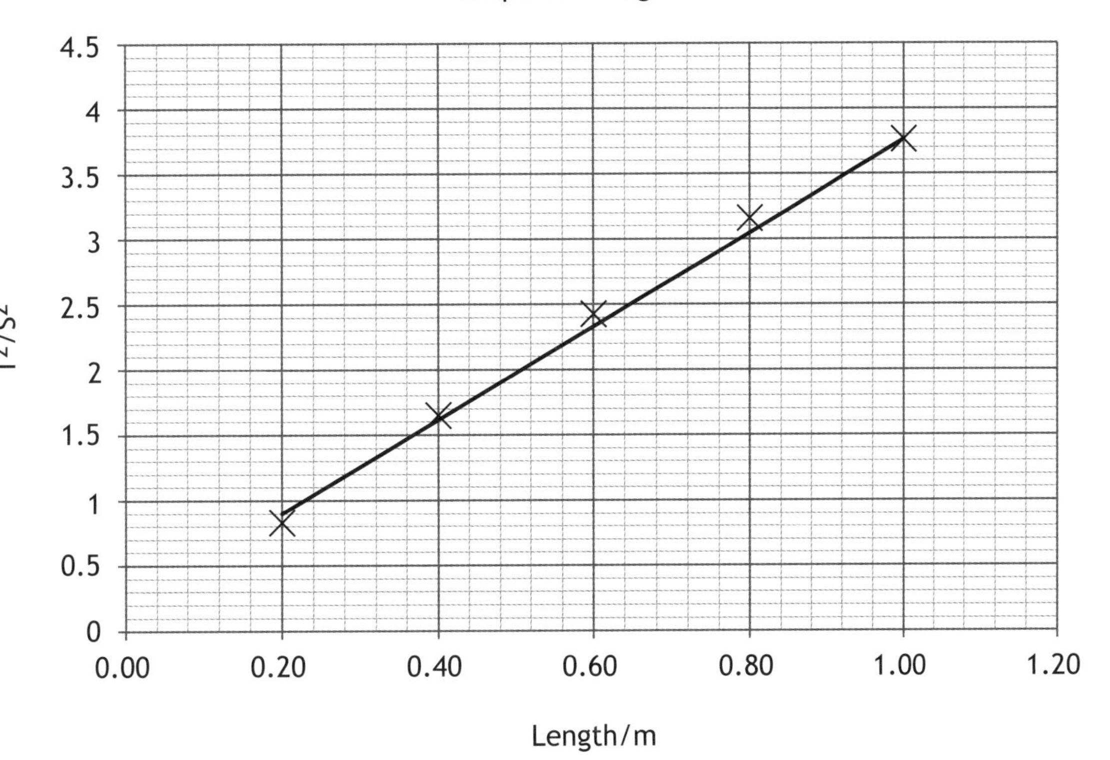


Figure 8

(a) Use the graph to calculate the gravitational field strength. **4**

MARKS | DO NOT WRITE IN THIS MARGIN

**8. (continued)**

Using the linest function, the uncertainty in the gradient is found to be

$$\Delta m = \pm 0{\cdot}12$$

(b) Calculate the absolute uncertainty in the value of $g$. **3**

(c) A teacher tells the pupil that she has not included error bars.

State the advantages of including error bars. **2**

(d) State how the absolute uncertainty would be estimated in

(i) $T^2$ **4**

(ii) length, $l$. **2**

MARKS | DO NOT WRITE IN THIS MARGIN

**9.** A transverse wave is described by the expression

$$y = 8{\cdot}0 \sin(12t - 0{\cdot}50x)$$

where $t$ is in seconds and $x$ and $y$ are in metres.

(a) For this wave, calculate the

(i) frequency **3**

(ii) wavelength. **2**

(b) (i) Calculate the phase difference, in radians, between the point at $x = 3{\cdot}0$ m and the point at $x = 4{\cdot}0$ m. **4**

(ii) Calculate the time for the wave to travel between these two points. **3**

(c) The wave is reflected and loses some energy.

State a possible equation for the reflected wave. **2**

MARKS DO NOT WRITE IN THIS MARGIN

**10.** The percentage of light reflected from glass can be greatly increased by using thin film technology where a material of greater refractive index than glass is deposited on the glass.

Figure 10 (not to scale) shows a light ray being reflected at both surfaces. The reflected rays can interfere constructively giving maximum reflection of light. The film has a thickness, *d*.

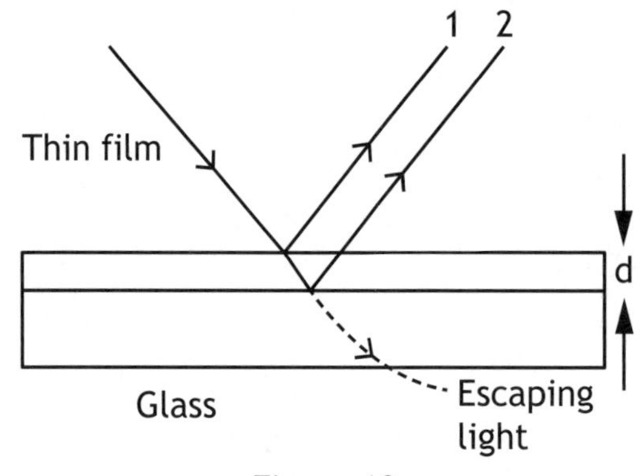


Figure 10

(a) State which principle of interference is behind this technology. **1**

(b) State what happens to the phase after reflection in:

(i) Ray 1 **1**

(ii) Ray 2 **1**

(c) Show, for normal incidence, the condition for maximum reflection is given by

$$d = \frac{\lambda}{4n}$$

where $n$ = refractive index of the film

$\lambda$ = wavelength of the light

$d$ = minimum thickness of the film **2**

MARKS DO NOT WRITE IN THIS MARGIN

**10. (continued)**

(d) A laser technician requires maximum reflection for a wavelength of 550 nm. The film material used is titanium dioxide which has a refractive index of 2·40.

Calculate the minimum required thickness. **2**

(e) The process in (d) produces a change from 4% to 70% reflectivity from the glass.

Suggest how the reflectivity could be increased to the required 99·9%. **2**

MARKS DO NOT WRITE IN THIS MARGIN

**11.** A student directs a white light source onto a modern television screen that is switched off. She notices that coloured fringes patterns are produced in the reflected light.

Use your knowledge of Physics to suggest an explanation for this observation. **3**

MARKS DO NOT WRITE IN THIS MARGIN

**12.** An alpha particle passes through a region that has perpendicular electric and magnetic fields, as shown in Figure 12A.

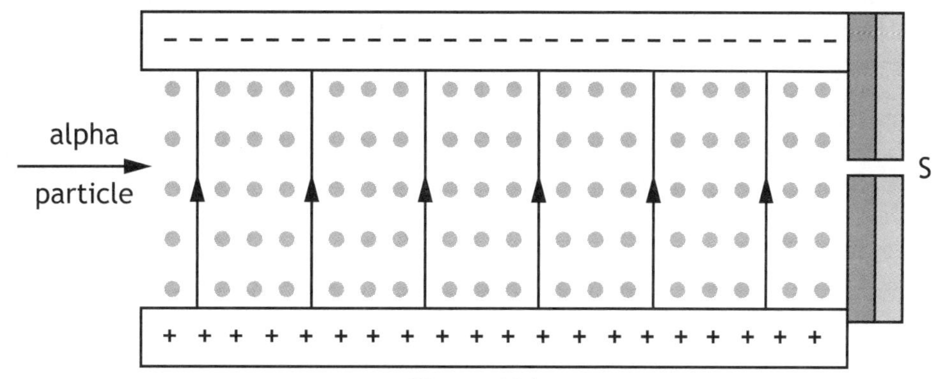


Figure 12A

The magnetic induction is 6·8 T and is directed out of the page.

The force on the alpha particle due to the magnetic field is $5{\cdot}0 \times 10^{-11}$ N.

(a) Show that the velocity of the alpha particle is $2{\cdot}3 \times 10^{7}$ m s$^{-1}$. **3**

(b) In order that the alpha particle exits through slit S, it must pass through the region undeflected.

Calculate the strength of the electric field that ensures the alpha particle exits through slit S. **3**

MARKS DO NOT WRITE IN THIS MARGIN

**12. (continued)**

(c) After passing through slit *S*, the alpha particle enters a region where there is the same uniform perpendicular magnetic field but no electric field as shown in Figure 12B.

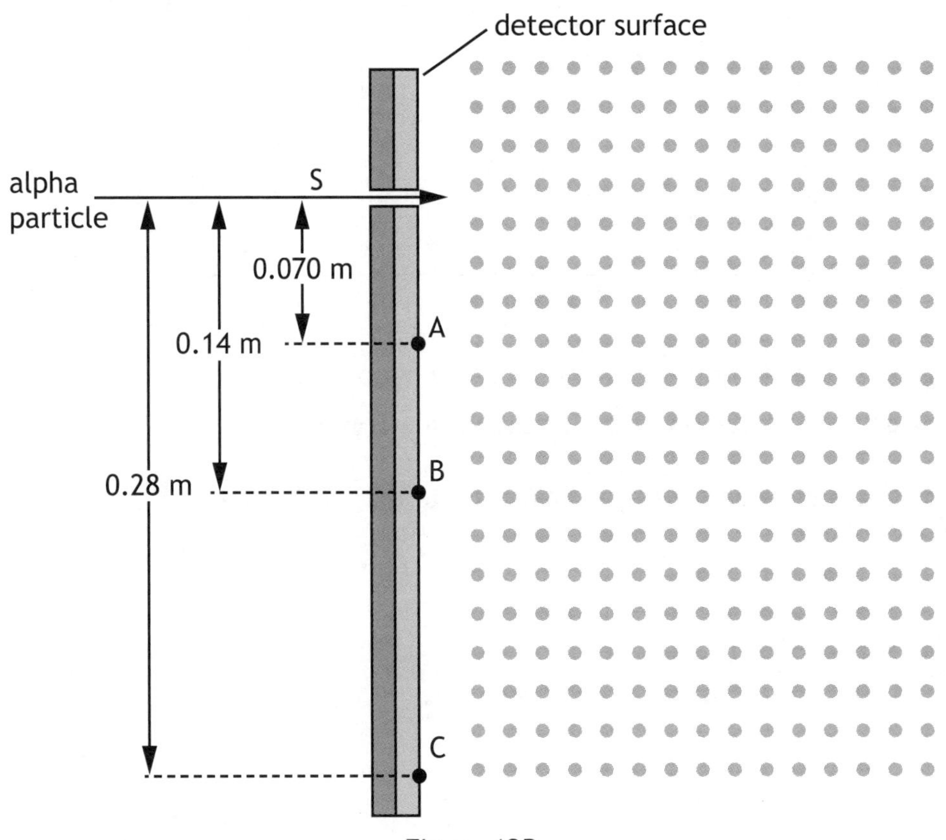


Figure 12B

This magnetic field causes the alpha particle to travel in a semi-circular path and hit the detector surface.

Points *A*, *B* and *C* are at distances of 0·070 m, 0·14 m and 0·28 m respectively from slit *S*.

Show, by calculation, which point on the detector surface is hit by the alpha particle. **4**

MARKS DO NOT WRITE IN THIS MARGIN

**12. (continued)**

(d) An electron travelling at the same speed as the alpha particle passes through slit S into the region of uniform magnetic field.

State two differences in the semi-circular path of the electron compared to the path of the alpha particle. Justify your answer. **3**

MARKS DO NOT WRITE IN THIS MARGIN

**13.** (a) A student investigates how the current in an inductor varies with the frequency of a voltage supply.

(i) Draw a suitable labelled circuit diagram of the apparatus required to carry out the investigation. **2**

(ii) The student collects the following data.

| Frequency/Hz | 40 | 60 | 80 | 100 | 120 |
|---|---|---|---|---|---|
| Current/mA | 148 | 101 | 76·0 | 58·2 | 50·0 |

Determine the relationship between the supply frequency and current for this inductor. **2**

(b) An inductor of inductance 3·0 H and negligible resistance is connected in a circuit with a 12 $\Omega$ resistor and supply voltage Vs as shown in Figure 13A.

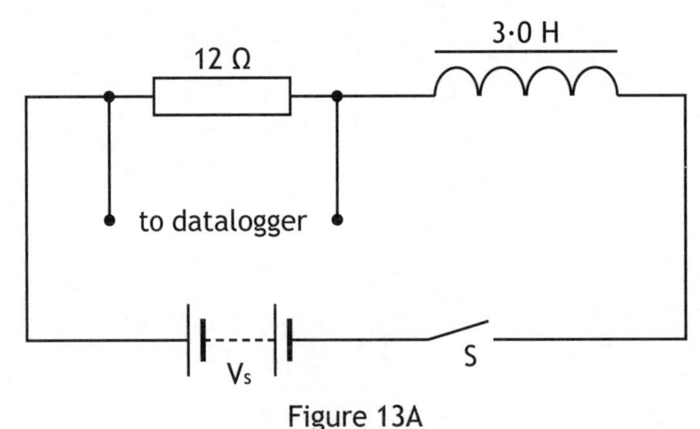


Figure 13A

The datalogger is set to calculate the back emf across the inductor. Switch S is initially open.

Switch S is now closed. Figure 12B shows how the back emf across the inductor varies from the instant the switch is closed.

MARKS DO NOT WRITE IN THIS MARGIN

**13. (continued)**

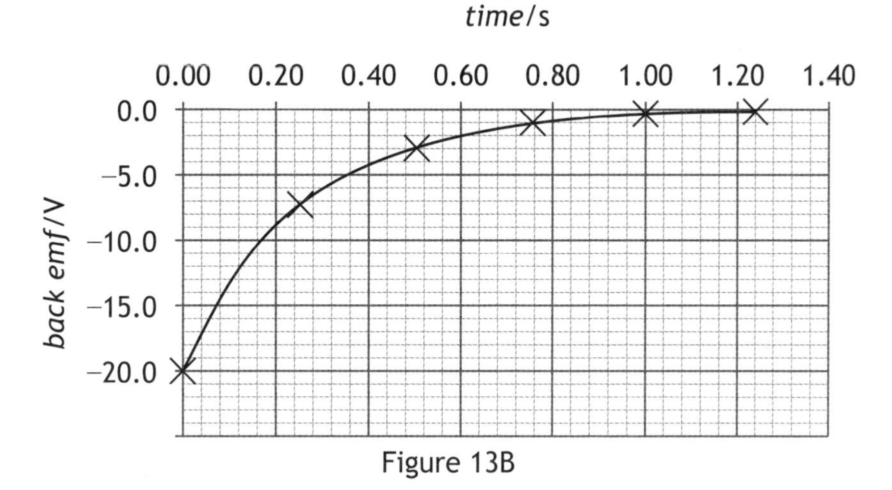


Figure 13B

(i) Determine the voltage across the resistor at $t = 0{\cdot}20$ s. **3**

(ii) Calculate the rate of change of current in the circuit at $t = 0{\cdot}40$ s. **3**

(iii) State why the magnitude of the back emf is greatest at $t = 0$. **1**

MARKS DO NOT WRITE IN THIS MARGIN

**13. (continued)**

(c) A tuned circuit consisting of an inductor, capacitor and resistor is shown in Figure 13C.

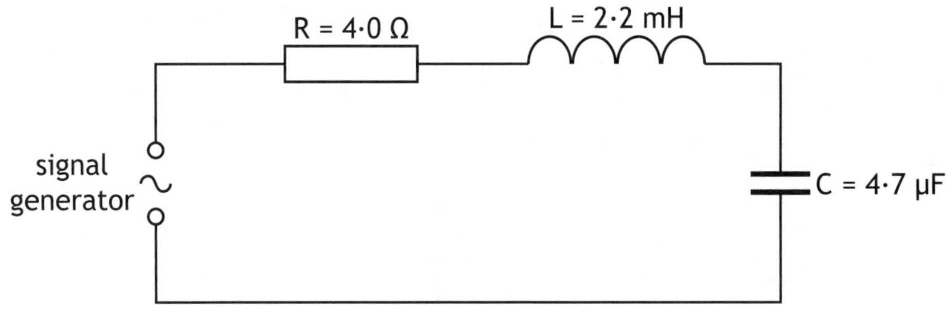


Figure 13C

The impedance $Z$, measured in ohms, of the circuit is given by the relationship

$$Z = \sqrt{R^2 + (X_L - X_C)^2}$$

where the symbols have their usual meanings.

(i) At a particular frequency $f_0$, the impedance of the circuit is a minimum. Show that $f_0$ is given by **1**

$$f_0 = \frac{1}{2\pi\sqrt{LC}}$$

(ii) Calculate the frequency $f_0$. **2**

(iii) State the minimum impedance of the circuit. **1**

**[END OF MODEL PAPER]**

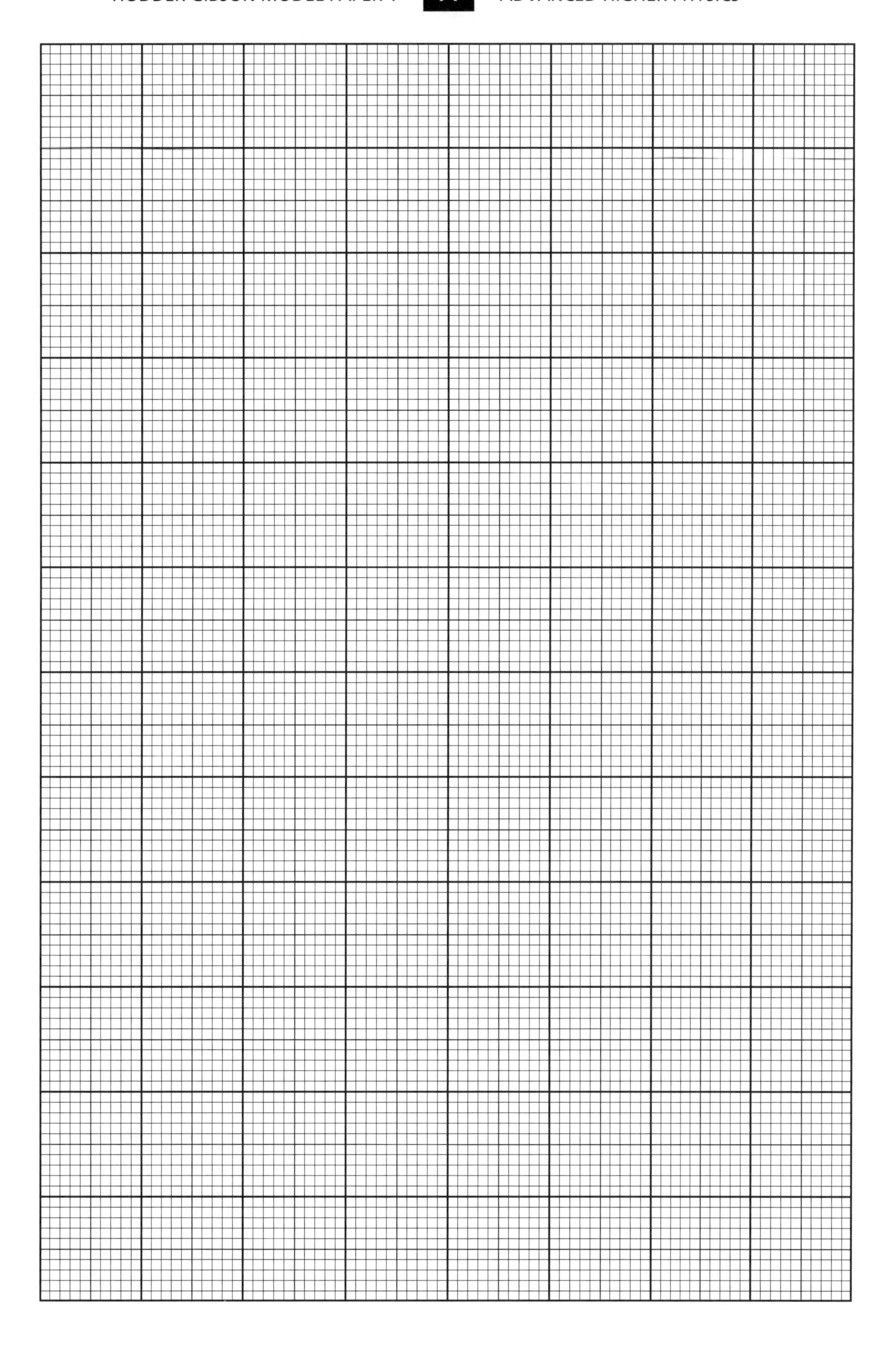

## ADDITIONAL SPACE FOR ANSWERS AND ROUGH WORK

ADVANCED HIGHER

# Model Paper 2

Whilst this Model Paper has been specially commissioned by Hodder Gibson for use as practice for the Advanced Higher (for Curriculum for Excellence) exams, the key reference document remains the SQA Specimen Paper 2015 and the SQA Exam Paper 2016.

FOR OFFICIAL USE

AH

National Qualifications MODEL PAPER 2

Mark

# Physics

Duration — 2 hours 30 minutes

**Fill in these boxes and read what is printed below.**

Full name of centre

Town

Forename(s)

Surname

Number of seat

Date of birth

Day Month Year

D D M M Y Y

Scottish candidate number

**Total marks — 140**

Attempt ALL questions.

Reference may be made to the Physics Relationships Sheet and the Data Sheet on *Page two*.

Write your answers clearly in the spaces provided in this booklet. Additional space for answers is provided at the end of this booklet. If you use this space you must clearly identify the question number you are attempting. Any rough work must be written in this booklet. You should score through your rough work when you have written your final copy.

Use **blue** or **black** ink.

Before leaving the examination room you must give this booklet to the Invigilator; if you do not, you may lose all the marks for this paper.

**DATA SHEET**

COMMON PHYSICAL QUANTITIES

| *Quantity* | *Symbol* | *Value* | *Quantity* | *Symbol* | *Value* |
|---|---|---|---|---|---|
| Gravitational acceleration on Earth | $g$ | $9{\cdot}8\,\text{m s}^{-2}$ | Mass of electron | $m_e$ | $9{\cdot}11 \times 10^{-31}\,\text{kg}$ |
| Radius of Earth | $R_E$ | $6{\cdot}4 \times 10^{6}\,\text{m}$ | Charge on electron | $e$ | $-1{\cdot}60 \times 10^{-19}\,\text{C}$ |
| Mass of Earth | $M_E$ | $6{\cdot}0 \times 10^{24}\,\text{kg}$ | Mass of neutron | $m_n$ | $1{\cdot}675 \times 10^{-27}\,\text{kg}$ |
| Mass of Moon | $M_M$ | $7{\cdot}3 \times 10^{22}\,\text{kg}$ | Mass of proton | $m_p$ | $1{\cdot}673 \times 10^{-27}\,\text{kg}$ |
| Radius of Moon | $R_M$ | $1{\cdot}7 \times 10^{6}\,\text{m}$ | Mass of alpha particle | $m_a$ | $6{\cdot}645 \times 10^{-27}\,\text{kg}$ |
| Mean Radius of Moon Orbit | | $3{\cdot}84 \times 10^{8}\,\text{m}$ | Charge on alpha particle | | $3{\cdot}20 \times 10^{-19}\,\text{C}$ |
| Solar radius | | $6{\cdot}955 \times 10^{8}\,\text{m}$ | Planck's constant | $h$ | $6{\cdot}63 \times 10^{-34}\,\text{J s}$ |
| Mass of Sun | | $2{\cdot}0 \times 10^{30}\,\text{kg}$ | | | |
| 1 AU | | $1{\cdot}5 \times 10^{11}\,\text{m}$ | Permittivity of free space | $\varepsilon_0$ | $8{\cdot}85 \times 10^{-12}\,\text{F m}^{-1}$ |
| Stefan-Boltzmann constant | $\sigma$ | $5{\cdot}67 \times 10^{-8}\,\text{W m}^{-2}\text{K}^{-4}$ | Permeability of free space | $\mu_0$ | $4\pi \times 10^{-7}\,\text{H m}^{-1}$ |
| Universal constant of gravitation | $G$ | $6{\cdot}67 \times 10^{-11}\,\text{m}^3\,\text{kg}^{-1}\,\text{s}^{-2}$ | Speed of light in vacuum | $c$ | $3{\cdot}0 \times 10^{8}\,\text{m s}^{-1}$ |
| | | | Speed of sound in air | $v$ | $3{\cdot}4 \times 10^{2}\,\text{m s}^{-1}$ |

REFRACTIVE INDICES

The refractive indices refer to sodium light of wavelength 589 nm and to substances at a temperature of 273 K.

| *Substance* | *Refractive index* | *Substance* | *Refractive index* |
|---|---|---|---|
| Diamond | 2·42 | Glycerol | 1·47 |
| Glass | 1·51 | Water | 1·33 |
| Ice | 1·31 | Air | 1·00 |
| Perspex | 1·49 | Magnesium Fluoride | 1·38 |

SPECTRAL LINES

| *Element* | *Wavelength*/nm | *Colour* | *Element* | *Wavelength*/nm | *Colour* |
|---|---|---|---|---|---|
| Hydrogen | 656 | Red | Cadmium | 644 | Red |
| | 486 | Blue-green | | 509 | Green |
| | 434 | Blue-violet | | 480 | Blue |
| | 410 | Violet | *Lasers* | | |
| | 397 | Ultraviolet | *Element* | *Wavelength*/nm | *Colour* |
| | 389 | Ultraviolet | Carbon dioxide | 9550 } 10590 | Infrared |
| Sodium | 589 | Yellow | Helium-neon | 633 | Red |

PROPERTIES OF SELECTED MATERIALS

| *Substance* | *Density*/ kg m$^{-3}$ | *Melting Point*/ K | *Boiling Point*/K | *Specific Heat Capacity*/ J kg$^{-1}$ K$^{-1}$ | *Specific Latent Heat of Fusion*/ J kg$^{-1}$ | *Specific Latent Heat of Vaporisation*/ J kg$^{-1}$ |
|---|---|---|---|---|---|---|
| Aluminium | $2{\cdot}70 \times 10^{3}$ | 933 | 2623 | $9{\cdot}02 \times 10^{2}$ | $3{\cdot}95 \times 10^{5}$ | .... |
| Copper | $8{\cdot}96 \times 10^{3}$ | 1357 | 2853 | $3{\cdot}86 \times 10^{2}$ | $2{\cdot}05 \times 10^{5}$ | .... |
| Glass | $2{\cdot}60 \times 10^{3}$ | 1400 | .... | $6{\cdot}70 \times 10^{2}$ | .... | .... |
| Ice | $9{\cdot}20 \times 10^{2}$ | 273 | .... | $2{\cdot}10 \times 10^{3}$ | $3{\cdot}34 \times 10^{5}$ | .... |
| Glycerol | $1{\cdot}26 \times 10^{3}$ | 291 | 563 | $2{\cdot}43 \times 10^{3}$ | $1{\cdot}81 \times 10^{5}$ | $8{\cdot}30 \times 10^{5}$ |
| Methanol | $7{\cdot}91 \times 10^{2}$ | 175 | 338 | $2{\cdot}52 \times 10^{3}$ | $9{\cdot}9 \times 10^{4}$ | $1{\cdot}12 \times 10^{6}$ |
| Sea Water | $1{\cdot}02 \times 10^{3}$ | 264 | 377 | $3{\cdot}93 \times 10^{3}$ | .... | .... |
| Water | $1{\cdot}00 \times 10^{3}$ | 273 | 373 | $4{\cdot}19 \times 10^{3}$ | $3{\cdot}34 \times 10^{5}$ | $2{\cdot}26 \times 10^{6}$ |
| Air | 1·29 | .... | .... | .... | .... | .... |
| Hydrogen | $9{\cdot}0 \times 10^{-2}$ | 14 | 20 | $1{\cdot}43 \times 10^{4}$ | .... | $4{\cdot}50 \times 10^{5}$ |
| Nitrogen | 1·25 | 63 | 77 | $1{\cdot}04 \times 10^{3}$ | .... | $2{\cdot}00 \times 10^{5}$ |
| Oxygen | 1·43 | 55 | 90 | $9{\cdot}18 \times 10^{2}$ | .... | $2{\cdot}40 \times 10^{4}$ |

The gas densities refer to a temperature of 273 K and a pressure of $1{\cdot}01 \times 10^{5}$ Pa.

**Total marks — 140 marks**

**Attempt ALL questions**

MARKS | DO NOT WRITE IN THIS MARGIN

1. A turntable, radius $r$, rotates with a constant angular velocity $\omega$ about an axis of rotation. Point $X$ on the circumference of the turntable is moving with a tangential speed $v$, as shown in Figure 1A.

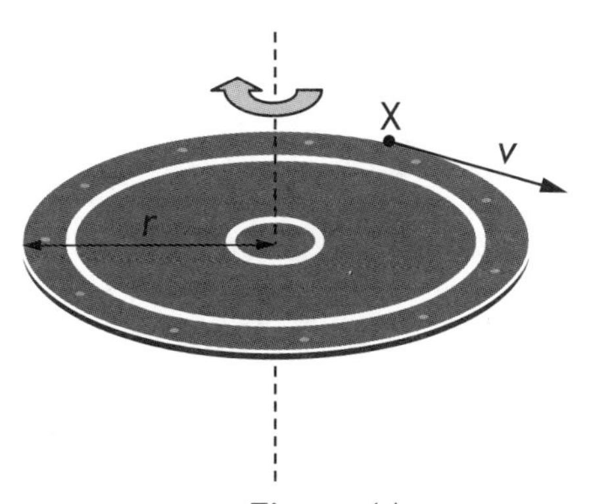


Figure 1A

Data recorded for the turntable is shown below.

| | |
|---|---|
| Angle of rotation | (3·1 ± 0·1) rad |
| Time taken for angle of rotation | (4·5 ± 0·1) s |
| Radius of disk | (0·148 ± 0·001) m |

(a) Calculate the tangential speed $v$. **3**

(b) Calculate the percentage uncertainty in this value of $v$. **4**

MARKS DO NOT WRITE IN THIS MARGIN

**2.** The front wheel of a racing bike can be considered to consist of five spokes and a rim, as shown in Figure 2A.

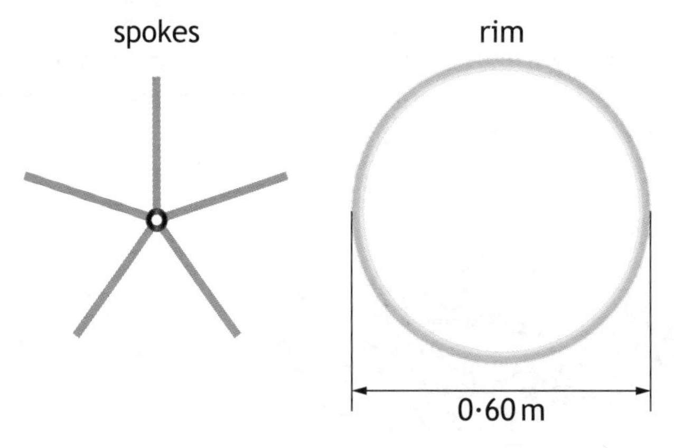


Figure 2A

The mass of each spoke is 0·040 kg and the mass of the rim is 0·24 kg. The wheel has a diameter of 0·60 m.

(a) (i) Each spoke can be considered as a uniform rod.

Calculate the moment of inertia of a spoke as the wheel rotates. **3**

(ii) Show that the total moment of inertia of the wheel is $2 \cdot 8 \times 10^{-2}$ kg m$^2$. **4**

MARKS DO NOT WRITE IN THIS MARGIN

**2. (continued)**

(b) The wheel is placed in a test rig and rotated as shown in Figure 2B.

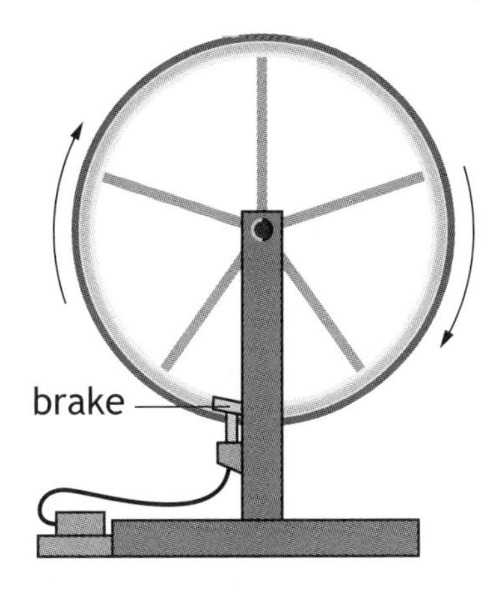


Figure 2B

(i) The tangential velocity of the rim is $19{\cdot}2\,\text{m}\,\text{s}^{-1}$. Calculate the angular velocity of the wheel. **3**

(ii) The brake is now applied to the rim of the wheel, bringing it uniformly to rest in $6{\cdot}7\,\text{s}$.

(A) Calculate the angular acceleration of the wheel. **3**

(B) Calculate the torque acting on the wheel. **3**

MARKS | DO NOT WRITE IN THIS MARGIN

**3.** (a) The gravitational field strength $g$ on the surface of Mars is $3{\cdot}7\,N\,kg^{-1}$.

The mass of Mars is $6{\cdot}4 \times 10^{23}\,kg$.

Show that the radius of Mars is $3{\cdot}4 \times 10^{6}\,m$. **3**

(b) (i) A satellite of mass $m$ has an orbit of radius $R$. Show that the angular velocity $\omega$ of the satellite is given by the expression

$$\omega = \sqrt{\frac{GM}{R^3}}$$

where the symbols have their usual meanings. **3**

(ii) A satellite remains above the same point on the equator of Mars as the planet spins on its axis.

Figure 3 shows this satellite orbiting at a height of $1{\cdot}7 \times 10^{7}\,m$ above the Martian surface.

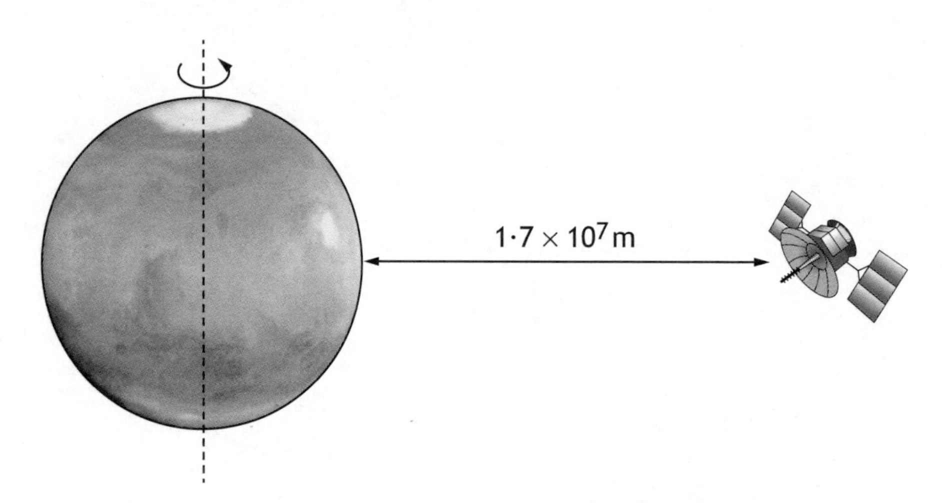


Figure 3 (not to scale)

MARKS DO NOT WRITE IN THIS MARGIN

**3. (b) (ii) (continued)**

Calculate the angular velocity of the satellite. 3

(iii) Calculate the length of one Martian day. 3

(c) The following table gives data about three planets orbiting the Sun.

| Planet | Radius $R$ of orbit around the Sun/$10^9$ m | Orbit period $T$ around the Sun/years |
|---|---|---|
| Venus | 108 | 0·62 |
| Mars | 227 | 1·88 |
| Jupiter | 780 | 12·0 |

Use **all** the data to show that $T^2$ is directly proportional to $R^3$ for these three planets. 3

MARKS DO NOT WRITE IN THIS MARGIN

**4.** A typical Hertzsprung—Russell (H—R) diagram is shown in Figure 4A.

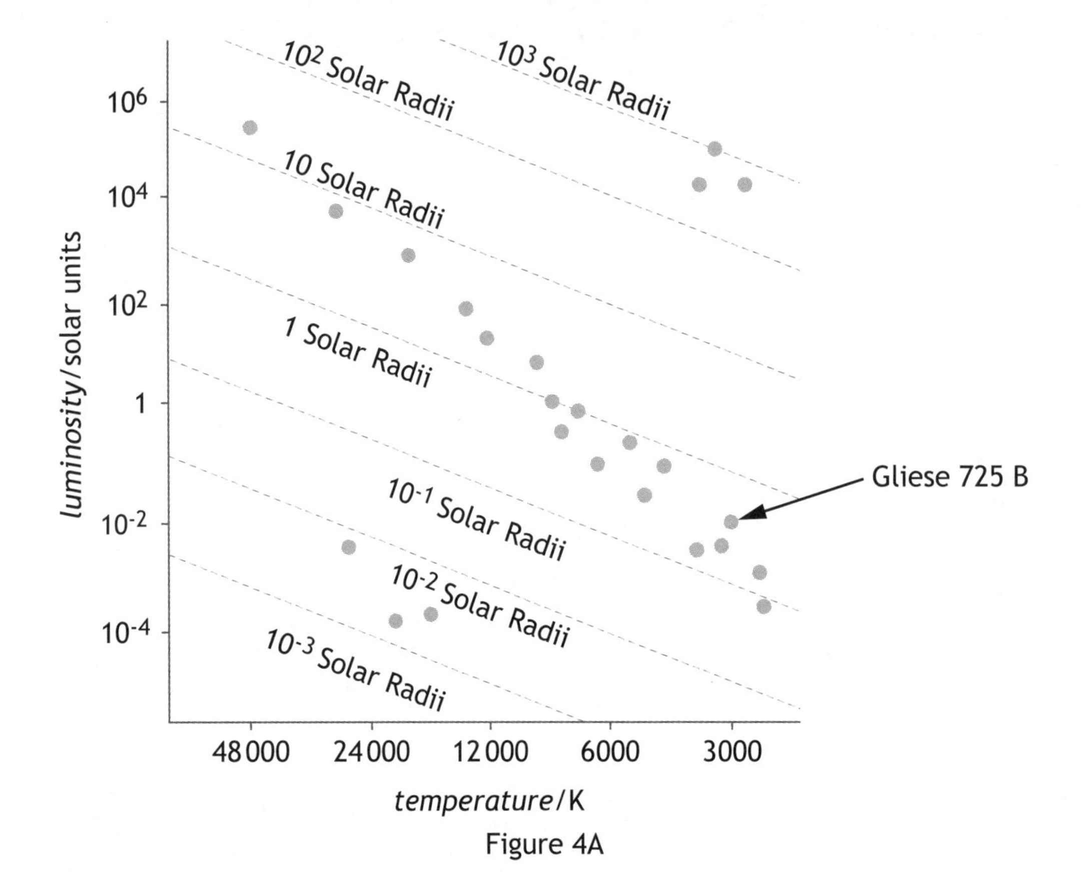


Figure 4A

(a) The luminosity of the Sun is $3{\cdot}9 \times 10^{26}$ W. Using information from Figure 4A:

(i) determine the luminosity in watts of Gliese 725 B. **2**

(ii) show that the radius of Gliese 725 B is $3 \times 10^8$ m. **3**

(iii) explain why it would be inappropriate to give the answer for part (ii) to more than one significant figure. **1**

MARKS DO NOT WRITE IN THIS MARGIN

**4. (continued)**

(b) Figure 4B shows how the radiation intensity varies with frequency for a black body radiator.

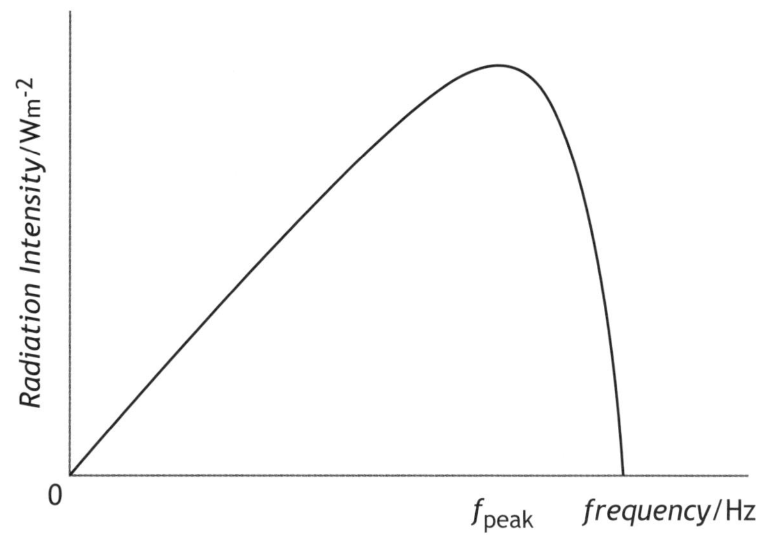


Figure 4B

This spectrum has a peak intensity at a frequency of $f_{peak}$.

$f_{peak}$ can be estimated using the relationship

$$f_{peak} = \frac{2 \cdot 8\, k_b T}{h}$$

where $k = 1 \cdot 38 \times 10^{-23}\ \text{JK}^{-1}$ (Boltzmann constant) and the other symbols have their usual meanings.

(i) Estimate $f_{peak}$ for Gliese 725 B. **3**

MARKS | DO NOT WRITE IN THIS MARGIN

**4. (b) (continued)**

(ii) The cosmic microwave background radiation (CMBR) has a spectrum which peaks at a wavelength of 1·9 mm.

Calculate the temperature of the CMBR. **4**

(c) Some astronomers have suggested that primordial black holes of mass $1{\cdot}0 \times 10^{-10}$ solar masses could make up the dark matter in our galaxy.

Determine the Schwarzchild radius of such a black hole. **3**

MARKS DO NOT WRITE IN THIS MARGIN

5. An engineer has to design a funfair ride that involves a "loop the loop".

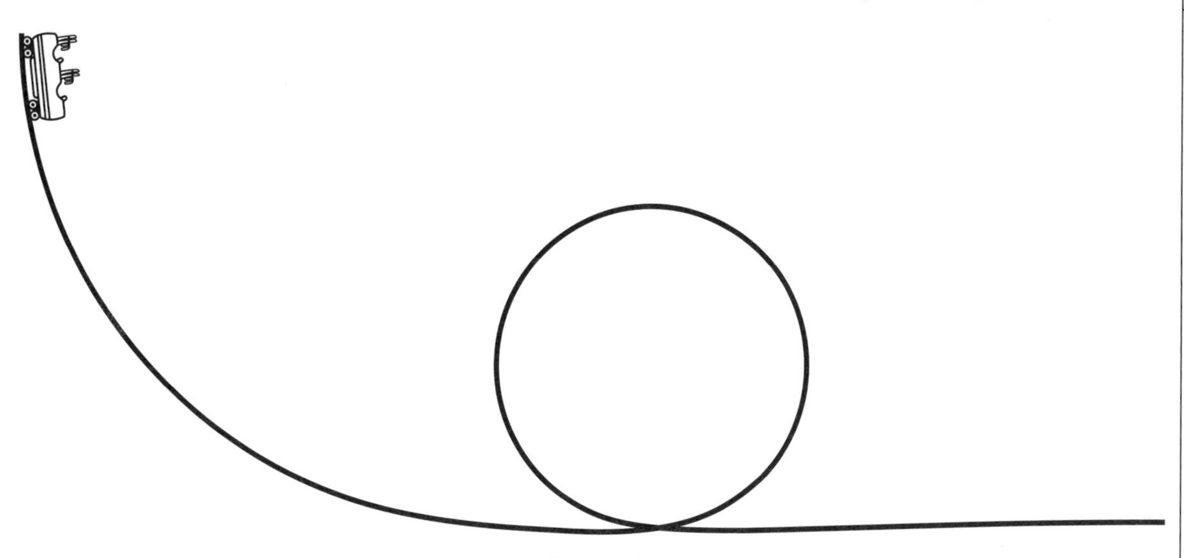

Figure 5

Using your knowledge of Physics, explain what factors she would have to take into account to ensure that the passengers are safe during the ride. **3**

MARKS | DO NOT WRITE IN THIS MARGIN

**6.** (a) The Bohr model of the hydrogen atom states that the angular momentum an electron is quantized.

(i) Calculate the minimum angular momentum of the electron in a hydrogen atom. **3**

(ii) When the electron has its minimum angular momentum it is in an orbit of radius $5{\cdot}3 \times 10^{-11}$ m. Calculate the linear momentum of the electron in this orbit. **3**

(iii) Calculate the de Broglie wavelength of the electron in this orbit. **3**

(b) One of the limitations of the model where electrons orbit a nucleus is that an orbiting electron is constantly accelerating and therefore should continuously emit electromagnetic radiation.

(i) What would happen to the orbit of the electron if electromagnetic radiation were to be continuously emitted? **1**

(ii) What is the name of the branch of physics that provides methods to determine the electron's position in terms of probability? **1**

MARKS DO NOT WRITE IN THIS MARGIN

**7.** A simple pendulum consists of a lead ball on the end of a long string as shown in Figure 7.

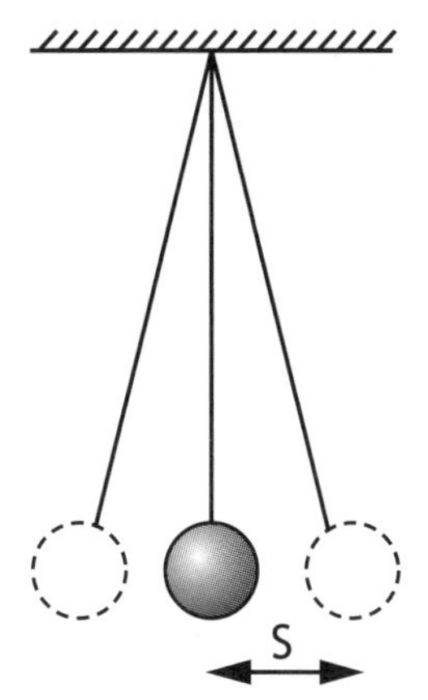


Figure 7

The ball moves with simple harmonic motion. At time $t$ the displacement $s$ of the ball is given by the expression

$$s = 2\cdot0 \times 10^{-2} \cos 4\cdot3t$$

where $s$ is in metres and $t$ in seconds.

(a) (i) State the definition of *simple harmonic motion*. **1**

(ii) Calculate the period of the pendulum. **3**

(b) Calculate the maximum speed of the ball. **3**

MARKS | DO NOT WRITE IN THIS MARGIN

**7. (continued)**

(c) The mass of the ball is $5 \cdot 0 \times 10^{-2}$ kg and the string has negligible mass.

Calculate the total energy of the pendulum. **3**

(d) The period $T$ of a pendulum is given by the expression

$$T = 2\pi\sqrt{\frac{L}{g}}$$

where $L$ is the length of the pendulum.

Calculate the length of this pendulum. **2**

(e) In the above case, the assumption has been made that the motion is not subject to *damping*. State what is meant by *damping*. **1**

MARKS DO NOT WRITE IN THIS MARGIN

**8.** A student uses a diffraction grating to find the wavelength of red light from a laser.

Figure 8 is produced by plotting sin $\theta$ against $m$ where $m$ is the order and $\theta$ the angular position of the bright fringes from the resulting interference pattern.

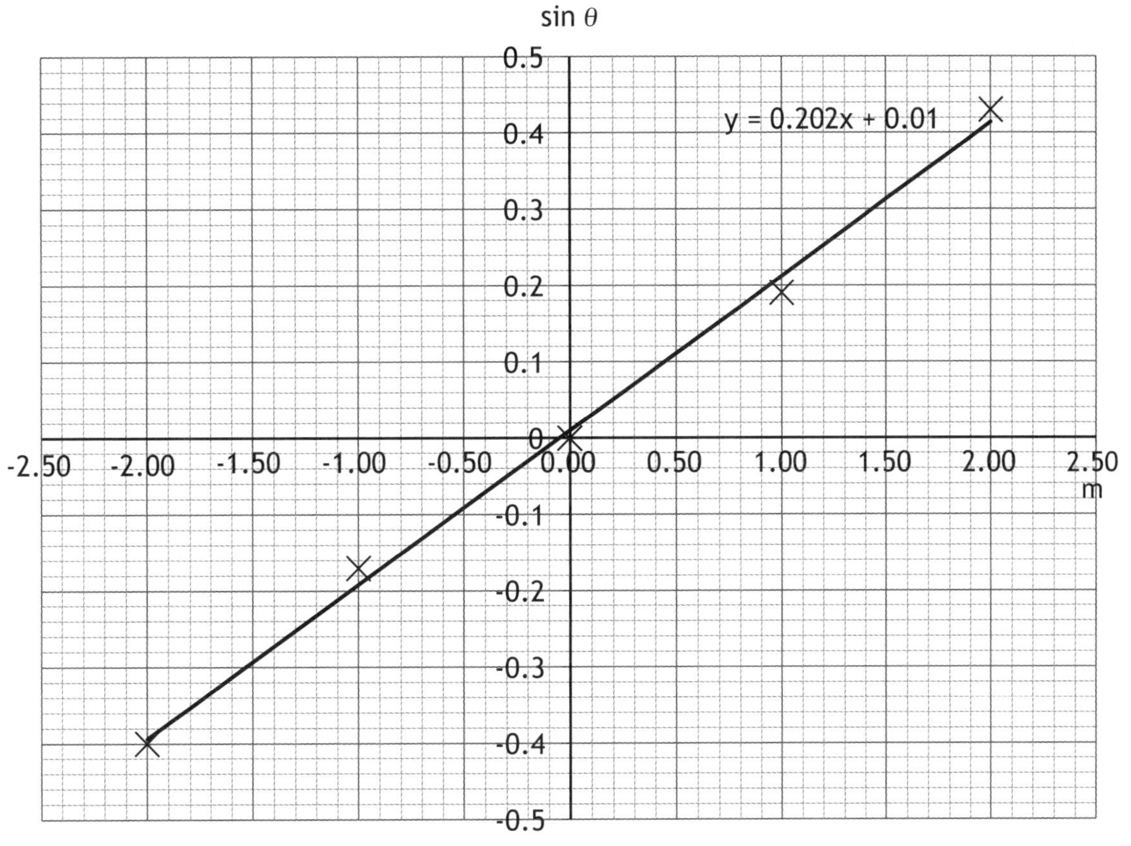


Figure 8

(a) Draw a labelled diagram of the experimental set up using a diffraction grating of 300 lines $mm^{-1}$, a screen and the laser. **2**

(b) Using a diagram, describe how $\theta$ is calculated for each value of $m$. **3**

MARKS | DO NOT WRITE IN THIS MARGIN

**8. (continued)**

(c) Using the graph, calculate the wavelength of the red light. **4**

It is found that the uncertainty in the gradient of the line is ± 0·0067.

(d) Calculate the absolute uncertainty in the calculated wavelength. **3**

(e) Suggest two improvements to the procedures. **2**

MARKS | DO NOT WRITE IN THIS MARGIN

**9.** The apparatus shown in Figure 9 is set up to measure the speed of transverse waves on a stretched string.

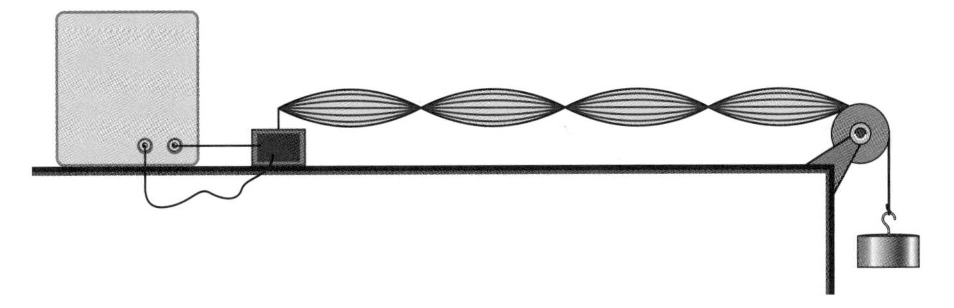

Figure 9

The following data are obtained.

Distance between adjacent nodes = (0·150 ± 0·005) m

Frequency of signal generator = (250 ± 10) Hz

(a) Show that the wave speed is 75 m s$^{-1}$. **3**

(b) Calculate the absolute uncertainty in this value for the wave speed. Express your answer in the form (75 ± ____ ) m s$^{-1}$. **5**

(c) (i) In an attempt to reduce the absolute uncertainty, the frequency of the signal generator is increased to (500 ± 10) Hz. Explain why this will **not** result in a reduced absolute uncertainty. **1**

(ii) State how the absolute uncertainty in wave speed could be reduced. **1**

MARKS DO NOT WRITE IN THIS MARGIN

**10.** In 1900, the British physicist Lord Kelvin is said to have pronounced:

*"There is nothing new to be discovered in physics now. All that remains is more and more precise measurement".*

Use your knowledge of Physics to comment on this statement. **3**

MARKS DO NOT WRITE IN THIS MARGIN

**11.** (a) Two point charges $Q_1$ and $Q_2$ each has a charge of $-4{\cdot}0\,\mu C$. The charges are 0·60 m apart as shown in Figure 11A.

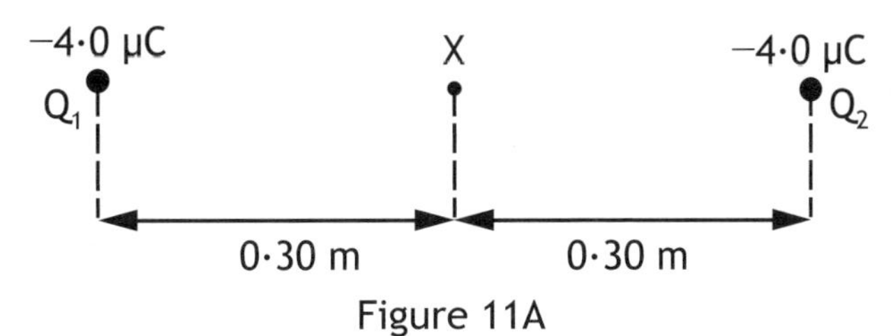


Figure 11A

(i) Draw a diagram to show the electric field lines between charges $Q_1$ and $Q_2$. **2**

(ii) Calculate the electrostatic potential at point *X*, midway between the charges. **3**

(b) A third point charge $Q_3$ is placed near the two charges as shown in Figure 11B.

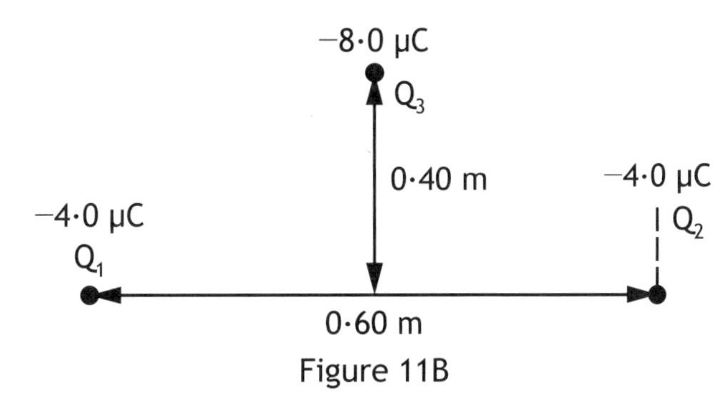


Figure 11B

(i) Show that the force between charges $Q_1$ and $Q_3$ is 1·2 N. **3**

(ii) Calculate the **magnitude** and **direction** of the resultant force on charge $Q_3$ due to charges $Q_1$ and $Q_2$. **3**

MARKS DO NOT WRITE IN THIS MARGIN

**12.** Identification of elements in a semiconductor sample can be carried out using an electron scanner to release atoms from the surface of the sample for analysis.

Electrons are accelerated from rest between a cathode and anode by a potential difference with each electron gaining $2{\cdot}4 \times 10^3$ eV.

A variable voltage supply connected to the deflection plates enables the beam to scan the sample between points A and B shown in Figure 12.

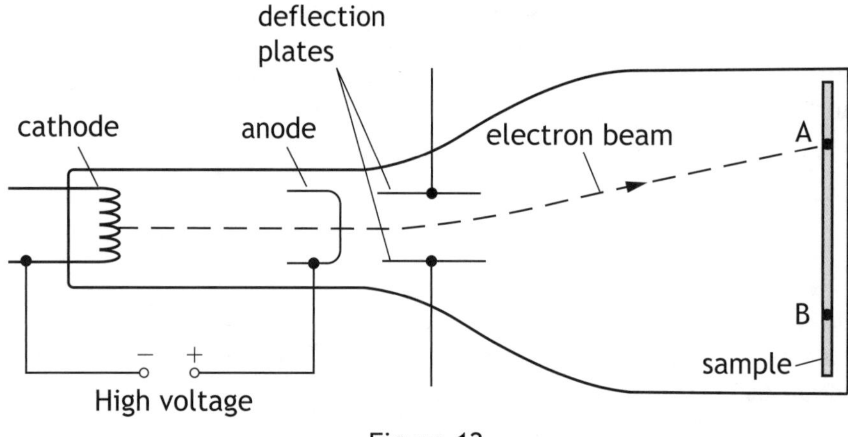


Figure 12

(a) Calculate the speed of the electrons as they pass through the anode. **3**

(b) Explain why the electron beam follows

(i) a curved path between the plates. **2**

(ii) a straight path beyond the plates. **1**

MARKS DO NOT WRITE IN THIS MARGIN

**12. (continued)**

When the potential difference across the deflection plates is 100V, the electron beam strikes the sample at position *A*.

(c) The deflection plates are 15·0 mm long and separated by 10·0 mm.

(i) Show that the vertical acceleration between the plates is $1{\cdot}76 \times 10^{15}\ \text{m s}^{-2}$. **3**

(ii) Calculate the vertical velocity of an electron as it emerges from between the plates. **4**

(d) The anode voltage is now increased. State what happens to the length of the sample scanned by the electron beam. You must justify your answer. **3**

MARKS DO NOT WRITE IN THIS MARGIN

**13.** A student is investigating the charging and discharging of a capacitor.

The circuit used is shown in Figure 13A.

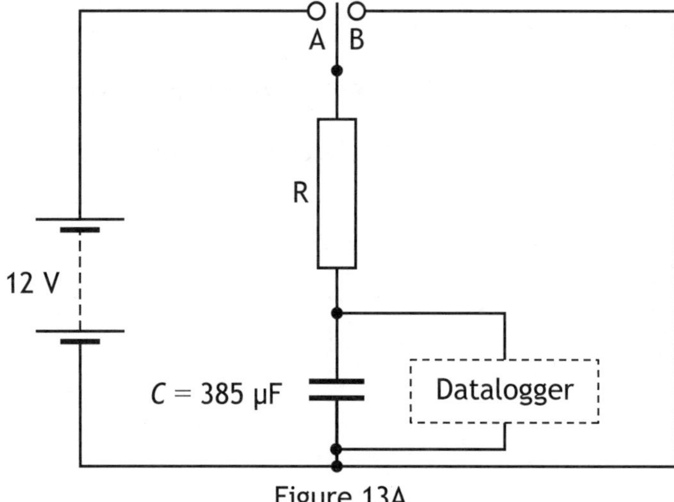


Figure 13A

With the switch in position *A*, the capacitor charges. To discharge the capacitor, the switch is moved to position *B*. The data logger monitors the voltage across the capacitor.

The graph in Figure 13B shows how the voltage across the capacitor changes during discharge.

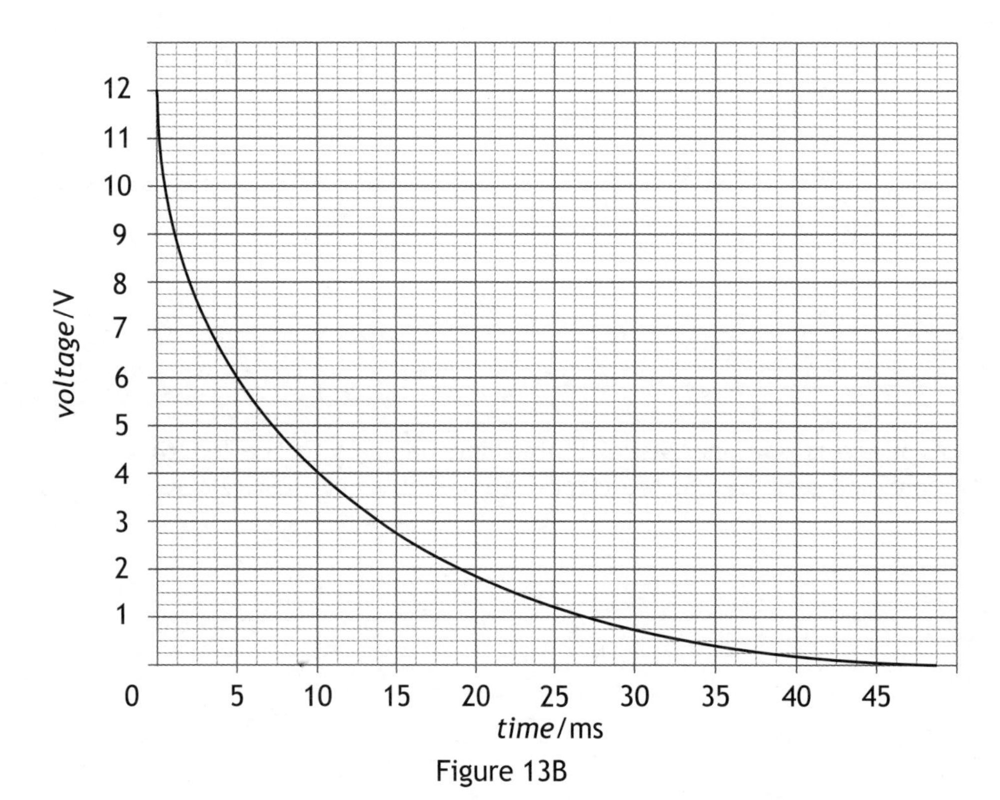


Figure 13B

MARKS | DO NOT WRITE IN THIS MARGIN

**13. (continued)**

(a) Determine the time constant from the graph. **2**

(b) Calculate the resistance of resistor *R*. **3**

**[END OF MODEL PAPER]**

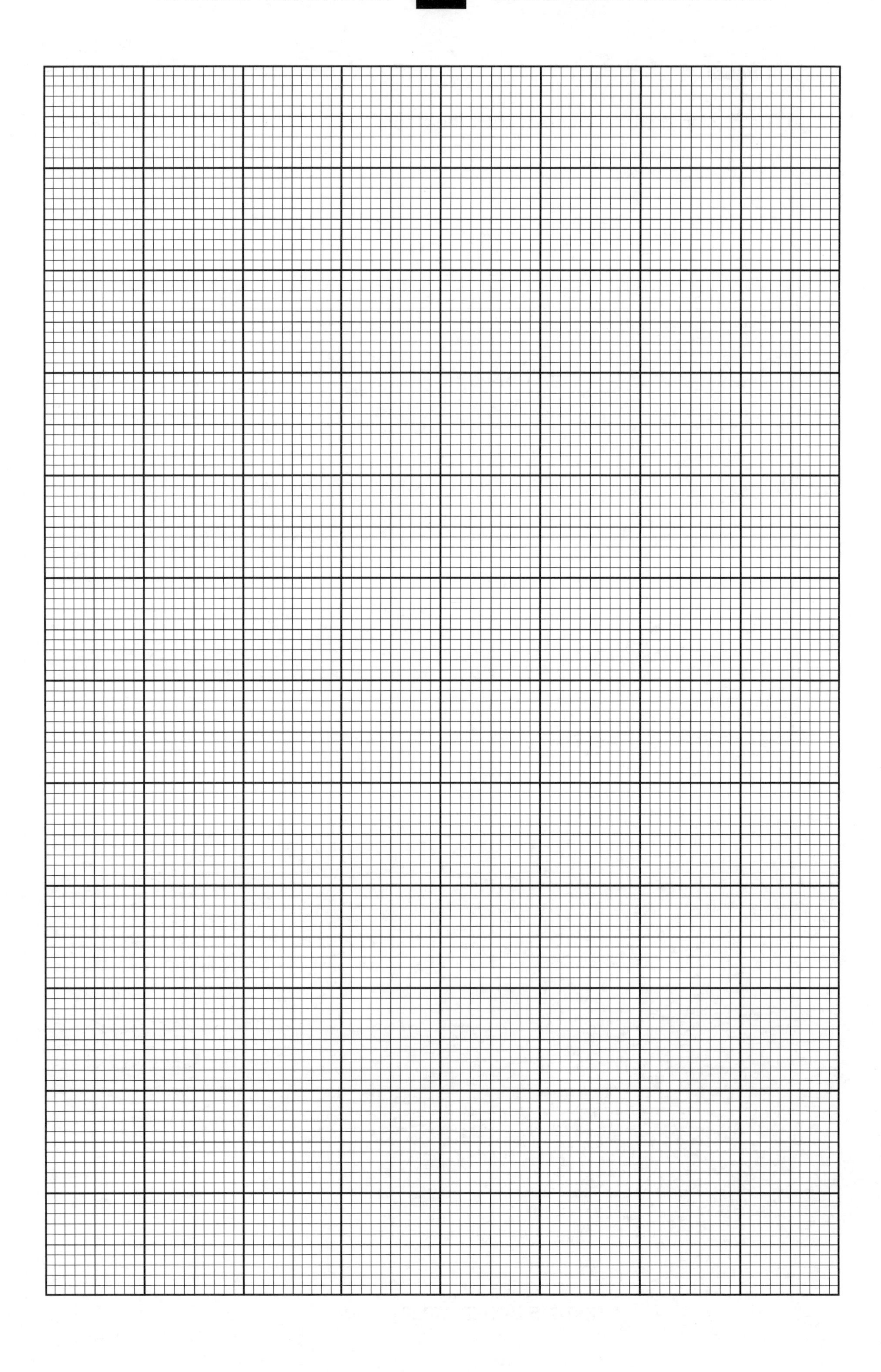

ADDITIONAL SPACE FOR ANSWERS AND ROUGH WORK

**[BLANK PAGE]**

**DO NOT WRITE ON THIS PAGE**

ADVANCED HIGHER

# 2016

National
Qualifications
2016

**X757/77/11**

# Physics
# Relationships Sheet

TUESDAY, 24 MAY
9:00 AM – 11:30 AM

# Relationships required for Physics Advanced Higher

$$v = \frac{ds}{dt}$$

$$a = \frac{dv}{dt} = \frac{d^2s}{dt^2}$$

$$v = u + at$$

$$s = ut + \frac{1}{2}at^2$$

$$v^2 = u^2 + 2as$$

$$\omega = \frac{d\theta}{dt}$$

$$\alpha = \frac{d\omega}{dt} = \frac{d^2\theta}{dt^2}$$

$$\omega = \omega_o + \alpha t$$

$$\theta = \omega_o t + \frac{1}{2}\alpha t^2$$

$$\omega^2 = \omega_o^2 + 2\alpha\theta$$

$$s = r\theta$$

$$v = r\omega$$

$$a_t = r\alpha$$

$$a_r = \frac{v^2}{r} = r\omega^2$$

$$F = \frac{mv^2}{r} = mr\omega^2$$

$$T = Fr$$

$$T = I\alpha$$

$$L = mvr = mr^2\omega$$

$$L = I\omega$$

$$E_K = \frac{1}{2}I\omega^2$$

$$F = G\frac{Mm}{r^2}$$

$$V = -\frac{GM}{r}$$

$$v = \sqrt{\frac{2GM}{r}}$$

$$\text{apparent brightness, } b = \frac{L}{4\pi r^2}$$

$$\text{Power per unit area} = \sigma T^4$$

$$L = 4\pi r^2\sigma T^4$$

$$r_{Schwarzschild} = \frac{2GM}{c^2}$$

$$E = hf$$

$$\lambda = \frac{h}{p}$$

$$mvr = \frac{nh}{2\pi}$$

$$\Delta x\,\Delta p_x \geq \frac{h}{4\pi}$$

$$\Delta E\,\Delta t \geq \frac{h}{4\pi}$$

$$F = qvB$$

$$\omega = 2\pi f$$

$$a = \frac{d^2y}{dt^2} = -\omega^2 y$$

$$y = A\cos\omega t \quad \text{or} \quad y = A\sin\omega t$$

$$v = \pm\omega\sqrt{(A^2 - y^2)}$$

$$E_K = \frac{1}{2}m\omega^2(A^2 - y^2)$$

$$E_P = \frac{1}{2}m\omega^2 y^2$$

$$y = A\sin 2\pi(ft - \frac{x}{\lambda})$$

$$E = kA^2$$

$$\phi = \frac{2\pi x}{\lambda}$$

$$\text{optical path difference} = m\lambda \quad \text{or} \quad \left(m + \frac{1}{2}\right)\lambda$$

where $m = 0, 1, 2....$

$$\Delta x = \frac{\lambda l}{2d}$$

$$d = \frac{\lambda}{4n}$$

$$\Delta x = \frac{\lambda D}{d}$$

$$n = \tan i_P$$

$$F = \frac{Q_1 Q_2}{4\pi\varepsilon_o r^2}$$

$$E = \frac{Q}{4\pi\varepsilon_o r^2}$$

$$V = \frac{Q}{4\pi\varepsilon_o r}$$

$$F = QE$$

$$V = Ed$$

$$F = IlB\sin\theta$$

$$B = \frac{\mu_o I}{2\pi r}$$

$$c = \frac{1}{\sqrt{\varepsilon_o \mu_o}}$$

$$t = RC$$

$$X_C = \frac{V}{I}$$

$$X_C = \frac{1}{2\pi fC}$$

$$\mathcal{E} = -L\frac{dI}{dt}$$

$$E = \frac{1}{2}LI^2$$

$$X_L = \frac{V}{I}$$

$$X_L = 2\pi fL$$

$$\frac{\Delta W}{W} = \sqrt{\left(\frac{\Delta X}{X}\right)^2 + \left(\frac{\Delta Y}{Y}\right)^2 + \left(\frac{\Delta Z}{Z}\right)^2}$$

$$\Delta W = \sqrt{\Delta X^2 + \Delta Y^2 + \Delta Z^2}$$

$d = \bar{v}t$

$s = \bar{v}t$

$v = u + at$

$s = ut + \frac{1}{2}at^2$

$v^2 = u^2 + 2as$

$s = \frac{1}{2}(u + v)t$

$W = mg$

$F = ma$

$E_W = Fd$

$E_P = mgh$

$E_K = \frac{1}{2}mv^2$

$P = \frac{E}{t}$

$p = mv$

$Ft = mv - mu$

$F = G\frac{Mm}{r^2}$

$t' = \frac{t}{\sqrt{1 - \left(\frac{v}{c}\right)^2}}$

$l' = l\sqrt{1 - \left(\frac{v}{c}\right)^2}$

$f_o = f_s\left(\frac{v}{v \pm v_s}\right)$

$z = \frac{\lambda_{observed} - \lambda_{rest}}{\lambda_{rest}}$

$z = \frac{v}{c}$

$v = H_0 d$

$E_W = QV$

$E = mc^2$

$E = hf$

$E_K = hf - hf_0$

$E_2 - E_1 = hf$

$T = \frac{1}{f}$

$v = f\lambda$

$d\sin\theta = m\lambda$

$n = \frac{\sin\theta_1}{\sin\theta_2}$

$\frac{\sin\theta_1}{\sin\theta_2} = \frac{\lambda_1}{\lambda_2} = \frac{v_1}{v_2}$

$\sin\theta_c = \frac{1}{n}$

$I = \frac{k}{d^2}$

$I = \frac{P}{A}$

path difference $= m\lambda$ or $\left(m + \frac{1}{2}\right)\lambda$ where $m = 0, 1, 2...$

random uncertainty $= \frac{\text{max. value - min. value}}{\text{number of values}}$

$V_{peak} = \sqrt{2}V_{rms}$

$I_{peak} = \sqrt{2}I_{rms}$

$Q = It$

$V = IR$

$P = IV = I^2R = \frac{V^2}{R}$

$R_T = R_1 + R_2 + ....$

$\frac{1}{R_T} = \frac{1}{R_1} + \frac{1}{R_2} + ....$

$E = V + Ir$

$V_1 = \left(\frac{R_1}{R_1 + R_2}\right)V_S$

$\frac{V_1}{V_2} = \frac{R_1}{R_2}$

$C = \frac{Q}{V}$

$E = \frac{1}{2}QV = \frac{1}{2}CV^2 = \frac{1}{2}\frac{Q^2}{C}$

# Additional Relationships

**Circle**

$circumference = 2\pi r$

$area = \pi r^2$

**Sphere**

$area = 4\pi r^2$

$volume = \frac{4}{3}\pi r^3$

**Trigonometry**

$\sin\theta = \frac{opposite}{hypotenuse}$

$\cos\theta = \frac{adjacent}{hypotenuse}$

$\tan\theta = \frac{opposite}{adjacent}$

$\sin^2\theta + \cos^2\theta = 1$

**Moment of inertia**

point mass
$I = mr^2$

rod about centre
$I = \frac{1}{12}ml^2$

rod about end
$I = \frac{1}{3}ml^2$

disc about centre
$I = \frac{1}{2}mr^2$

sphere about centre
$I = \frac{2}{5}mr^2$

**Table of standard derivatives**

| $f(x)$ | $f'(x)$ |
|---|---|
| $\sin ax$ | $a\cos ax$ |
| $\cos ax$ | $-a\sin ax$ |

**Table of standard integrals**

| $f(x)$ | $\int f(x)dx$ |
|---|---|
| $\sin ax$ | $-\frac{1}{a}\cos ax + C$ |
| $\cos ax$ | $\frac{1}{a}\sin ax + C$ |

# Electron Arrangements of Elements

**Key**

| Atomic number<br>Symbol<br>Electron arrangement<br>Name |
|---|

## Transition Elements

| Group 1 (1) | Group 2 (2) | (3) | (4) | (5) | (6) | (7) | (8) | (9) | (10) | (11) | (12) | Group 3 (13) | Group 4 (14) | Group 5 (15) | Group 6 (16) | Group 7 (17) | Group 0 (18) |
|---|---|---|---|---|---|---|---|---|---|---|---|---|---|---|---|---|---|
| 1<br>H<br>1<br>Hydrogen | | | | | | | | | | | | | | | | | 2<br>He<br>2<br>Helium |
| 3<br>Li<br>2, 1<br>Lithium | 4<br>Be<br>2, 2<br>Beryllium | | | | | | | | | | | 5<br>B<br>2, 3<br>Boron | 6<br>C<br>2, 4<br>Carbon | 7<br>N<br>2, 5<br>Nitrogen | 8<br>O<br>2, 6<br>Oxygen | 9<br>F<br>2, 7<br>Fluorine | 10<br>Ne<br>2, 8<br>Neon |
| 11<br>Na<br>2, 8, 1<br>Sodium | 12<br>Mg<br>2, 8, 2<br>Magnesium | | | | | | | | | | | 13<br>Al<br>2, 8, 3<br>Aluminium | 14<br>Si<br>2, 8, 4<br>Silicon | 15<br>P<br>2, 8, 5<br>Phosphorus | 16<br>S<br>2, 8, 6<br>Sulphur | 17<br>Cl<br>2, 8, 7<br>Chlorine | 18<br>Ar<br>2, 8, 8<br>Argon |
| 19<br>K<br>2, 8, 8, 1<br>Potassium | 20<br>Ca<br>2, 8, 8, 2<br>Calcium | 21<br>Sc<br>2, 8, 9, 2<br>Scandium | 22<br>Ti<br>2, 8, 10, 2<br>Titanium | 23<br>V<br>2, 8, 11, 2<br>Vanadium | 24<br>Cr<br>2, 8, 13, 1<br>Chromium | 25<br>Mn<br>2, 8, 13, 2<br>Manganese | 26<br>Fe<br>2, 8, 14, 2<br>Iron | 27<br>Co<br>2, 8, 15, 2<br>Cobalt | 28<br>Ni<br>2, 8, 16, 2<br>Nickel | 29<br>Cu<br>2, 8, 18, 1<br>Copper | 30<br>Zn<br>2, 8, 18, 2<br>Zinc | 31<br>Ga<br>2, 8, 18, 3<br>Gallium | 32<br>Ge<br>2, 8, 18, 4<br>Germanium | 33<br>As<br>2, 8, 18, 5<br>Arsenic | 34<br>Se<br>2, 8, 18, 6<br>Selenium | 35<br>Br<br>2, 8, 18, 7<br>Bromine | 36<br>Kr<br>2, 8, 18, 8<br>Krypton |
| 37<br>Rb<br>2, 8, 18, 8, 1<br>Rubidium | 38<br>Sr<br>2, 8, 18, 8, 2<br>Strontium | 39<br>Y<br>2, 8, 18, 9, 2<br>Yttrium | 40<br>Zr<br>2, 8, 18, 10, 2<br>Zirconium | 41<br>Nb<br>2, 8, 18, 12, 1<br>Niobium | 42<br>Mo<br>2, 8, 18, 13, 1<br>Molybdenum | 43<br>Tc<br>2, 8, 18, 13, 2<br>Technetium | 44<br>Ru<br>2, 8, 18, 15, 1<br>Ruthenium | 45<br>Rh<br>2, 8, 18, 16, 1<br>Rhodium | 46<br>Pd<br>2, 8, 18, 18, 0<br>Palladium | 47<br>Ag<br>2, 8, 18, 18, 1<br>Silver | 48<br>Cd<br>2, 8, 18, 18, 2<br>Cadmium | 49<br>In<br>2, 8, 18, 18, 3<br>Indium | 50<br>Sn<br>2, 8, 18, 18, 4<br>Tin | 51<br>Sb<br>2, 8, 18, 18, 5<br>Antimony | 52<br>Te<br>2, 8, 18, 18, 6<br>Tellurium | 53<br>I<br>2, 8, 18, 18, 7<br>Iodine | 54<br>Xe<br>2, 8, 18, 18, 8<br>Xenon |
| 55<br>Cs<br>2, 8, 18, 18, 8, 1<br>Caesium | 56<br>Ba<br>2, 8, 18, 18, 8, 2<br>Barium | 57<br>La<br>2, 8, 18, 18, 9, 2<br>Lanthanum | 72<br>Hf<br>2, 8, 18, 32, 10, 2<br>Hafnium | 73<br>Ta<br>2, 8, 18, 32, 11, 2<br>Tantalum | 74<br>W<br>2, 8, 18, 32, 12, 2<br>Tungsten | 75<br>Re<br>2, 8, 18, 32, 13, 2<br>Rhenium | 76<br>Os<br>2, 8, 18, 32, 14, 2<br>Osmium | 77<br>Ir<br>2, 8, 18, 32, 15, 2<br>Iridium | 78<br>Pt<br>2, 8, 18, 32, 17, 1<br>Platinum | 79<br>Au<br>2, 8, 18, 32, 18, 1<br>Gold | 80<br>Hg<br>2, 8, 18, 32, 18, 2<br>Mercury | 81<br>Tl<br>2, 8, 18, 32, 18, 3<br>Thallium | 82<br>Pb<br>2, 8, 18, 32, 18, 4<br>Lead | 83<br>Bi<br>2, 8, 18, 32, 18, 5<br>Bismuth | 84<br>Po<br>2, 8, 18, 32, 18, 6<br>Polonium | 85<br>At<br>2, 8, 18, 32, 18, 7<br>Astatine | 86<br>Rn<br>2, 8, 18, 32, 18, 8<br>Radon |
| 87<br>Fr<br>2, 8, 18, 32, 18, 8, 1<br>Francium | 88<br>Ra<br>2, 8, 18, 32, 18, 8, 2<br>Radium | 89<br>Ac<br>2, 8, 18, 32, 18, 9, 2<br>Actinium | 104<br>Rf<br>2, 8, 18, 32, 32, 10, 2<br>Rutherfordium | 105<br>Db<br>2, 8, 18, 32, 32, 11, 2<br>Dubnium | 106<br>Sg<br>2, 8, 18, 32, 32, 12, 2<br>Seaborgium | 107<br>Bh<br>2, 8, 18, 32, 32, 13, 2<br>Bohrium | 108<br>Hs<br>2, 8, 18, 32, 32, 14, 2<br>Hassium | 109<br>Mt<br>2, 8, 18, 32, 32, 15, 2<br>Meitnerium | | | | | | | | | |

| | | | | | | | | | | | | | | | |
|---|---|---|---|---|---|---|---|---|---|---|---|---|---|---|---|
| **Lanthanides** | 57<br>La<br>2, 8, 18, 18, 9, 2<br>Lanthanum | 58<br>Ce<br>2, 8, 18, 20, 8, 2<br>Cerium | 59<br>Pr<br>2, 8, 18, 21, 8, 2<br>Praseodymium | 60<br>Nd<br>2, 8, 18, 22, 8, 2<br>Neodymium | 61<br>Pm<br>2, 8, 18, 23, 8, 2<br>Promethium | 62<br>Sm<br>2, 8, 18, 24, 8, 2<br>Samarium | 63<br>Eu<br>2, 8, 18, 25, 8, 2<br>Europium | 64<br>Gd<br>2, 8, 18, 25, 9, 2<br>Gadolinium | 65<br>Tb<br>2, 8, 18, 27, 8, 2<br>Terbium | 66<br>Dy<br>2, 8, 18, 28, 8, 2<br>Dysprosium | 67<br>Ho<br>2, 8, 18, 29, 8, 2<br>Holmium | 68<br>Er<br>2, 8, 18, 30, 8, 2<br>Erbium | 69<br>Tm<br>2, 8, 18, 31, 8, 2<br>Thulium | 70<br>Yb<br>2, 8, 18, 32, 8, 2<br>Ytterbium | 71<br>Lu<br>2, 8, 18, 32, 9, 2<br>Lutetium |
| **Actinides** | 89<br>Ac<br>2, 8, 18, 32, 18, 9, 2<br>Actinium | 90<br>Th<br>2, 8, 18, 32, 18, 10, 2<br>Thorium | 91<br>Pa<br>2, 8, 18, 32, 20, 9, 2<br>Protactinium | 92<br>U<br>2, 8, 18, 32, 21, 9, 2<br>Uranium | 93<br>Np<br>2, 8, 18, 32, 22, 9, 2<br>Neptunium | 94<br>Pu<br>2, 8, 18, 32, 24, 8, 2<br>Plutonium | 95<br>Am<br>2, 8, 18, 32, 25, 8, 2<br>Americium | 96<br>Cm<br>2, 8, 18, 32, 25, 9, 2<br>Curium | 97<br>Bk<br>2, 8, 18, 32, 27, 8, 2<br>Berkelium | 98<br>Cf<br>2, 8, 18, 32, 28, 8, 2<br>Californium | 99<br>Es<br>2, 8, 18, 32, 29, 8, 2<br>Einsteinium | 100<br>Fm<br>2, 8, 18, 32, 30, 8, 2<br>Fermium | 101<br>Md<br>2, 8, 18, 32, 31, 8, 2<br>Mendelevium | 102<br>No<br>2, 8, 18, 32, 32, 8, 2<br>Nobelium | 103<br>Lr<br>2, 8, 18, 32, 32, 9, 2<br>Lawrencium |

FOR OFFICIAL USE

AH

National Qualifications 2016

Mark

# X757/77/01

# Physics

TUESDAY, 24 MAY

9:00 AM – 11:30 AM

**Fill in these boxes and read what is printed below.**

Full name of centre

Town

Forename(s)

Surname

Number of seat

Date of birth

Day Month Year

Scottish candidate number

**Total marks — 140**

Attempt ALL questions.

Reference may be made to the Physics Relationships Sheet X757/77/11 and the Data Sheet on *Page two*.

Write your answers clearly in the spaces provided in this booklet. Additional space for answers and rough work is provided at the end of this booklet. If you use this space you must clearly identify the question number you are attempting. Any rough work must be written in this booklet. You should score through your rough work when you have written your final copy.

Care should be taken to give an appropriate number of significant figures in the final answers to calculations.

Use **blue** or **black** ink.

Before leaving the examination room you must give this booklet to the Invigilator; if you do not, you may lose all the marks for this paper.

**DATA SHEET**

COMMON PHYSICAL QUANTITIES

| *Quantity* | *Symbol* | *Value* | *Quantity* | *Symbol* | *Value* |
|---|---|---|---|---|---|
| Gravitational acceleration on Earth | $g$ | $9{\cdot}8\,\text{m}\,\text{s}^{-2}$ | Mass of electron | $m_e$ | $9{\cdot}11 \times 10^{-31}\,\text{kg}$ |
| Radius of Earth | $R_E$ | $6{\cdot}4 \times 10^{6}\,\text{m}$ | Charge on electron | $e$ | $-1{\cdot}60 \times 10^{-19}\,\text{C}$ |
| Mass of Earth | $M_E$ | $6{\cdot}0 \times 10^{24}\,\text{kg}$ | Mass of neutron | $m_n$ | $1{\cdot}675 \times 10^{-27}\,\text{kg}$ |
| Mass of Moon | $M_M$ | $7{\cdot}3 \times 10^{22}\,\text{kg}$ | Mass of proton | $m_p$ | $1{\cdot}673 \times 10^{-27}\,\text{kg}$ |
| Radius of Moon | $R_M$ | $1{\cdot}7 \times 10^{6}\,\text{m}$ | Mass of alpha particle | $m_a$ | $6{\cdot}645 \times 10^{-27}\,\text{kg}$ |
| Mean Radius of Moon Orbit | | $3{\cdot}84 \times 10^{8}\,\text{m}$ | Charge on alpha particle | | $3{\cdot}20 \times 10^{-19}\,\text{C}$ |
| Solar radius | | $6{\cdot}955 \times 10^{8}\,\text{m}$ | Planck's constant | $h$ | $6{\cdot}63 \times 10^{-34}\,\text{J}\,\text{s}$ |
| Mass of Sun | | $2{\cdot}0 \times 10^{30}\,\text{kg}$ | Permittivity of free space | $\varepsilon_0$ | $8{\cdot}85 \times 10^{-12}\,\text{F}\,\text{m}^{-1}$ |
| 1 AU | | $1{\cdot}5 \times 10^{11}\,\text{m}$ | Permeability of free space | $\mu_0$ | $4\pi \times 10^{-7}\,\text{H}\,\text{m}^{-1}$ |
| Stefan-Boltzmann constant | $\sigma$ | $5{\cdot}67 \times 10^{-8}\,\text{W}\,\text{m}^{-2}\text{K}^{-4}$ | Speed of light in vacuum | $c$ | $3{\cdot}0 \times 10^{8}\,\text{m}\,\text{s}^{-1}$ |
| Universal constant of gravitation | $G$ | $6{\cdot}67 \times 10^{-11}\,\text{m}^3\,\text{kg}^{-1}\,\text{s}^{-2}$ | Speed of sound in air | $v$ | $3{\cdot}4 \times 10^{2}\,\text{m}\,\text{s}^{-1}$ |

REFRACTIVE INDICES

The refractive indices refer to sodium light of wavelength 589 nm and to substances at a temperature of 273 K.

| *Substance* | *Refractive index* | *Substance* | *Refractive index* |
|---|---|---|---|
| Diamond | 2·42 | Glycerol | 1·47 |
| Glass | 1·51 | Water | 1·33 |
| Ice | 1·31 | Air | 1·00 |
| Perspex | 1·49 | Magnesium Fluoride | 1·38 |

SPECTRAL LINES

| *Element* | *Wavelength*/nm | *Colour* | *Element* | *Wavelength*/nm | *Colour* |
|---|---|---|---|---|---|
| Hydrogen | 656 | Red | Cadmium | 644 | Red |
| | 486 | Blue-green | | 509 | Green |
| | 434 | Blue-violet | | 480 | Blue |
| | 410 | Violet | | *Lasers* | |
| | 397 | Ultraviolet | *Element* | *Wavelength*/nm | *Colour* |
| | 389 | Ultraviolet | Carbon dioxide | 9550, 10590 | Infrared |
| Sodium | 589 | Yellow | Helium-neon | 633 | Red |

PROPERTIES OF SELECTED MATERIALS

| *Substance* | *Density*/ $\text{kg}\,\text{m}^{-3}$ | *Melting Point*/ K | *Boiling Point*/K | *Specific Heat Capacity*/ $\text{J}\,\text{kg}^{-1}\,\text{K}^{-1}$ | *Specific Latent Heat of Fusion*/ $\text{J}\,\text{kg}^{-1}$ | *Specific Latent Heat of Vaporisation*/ $\text{J}\,\text{kg}^{-1}$ |
|---|---|---|---|---|---|---|
| Aluminium | $2{\cdot}70 \times 10^{3}$ | 933 | 2623 | $9{\cdot}02 \times 10^{2}$ | $3{\cdot}95 \times 10^{5}$ | .... |
| Copper | $8{\cdot}96 \times 10^{3}$ | 1357 | 2853 | $3{\cdot}86 \times 10^{2}$ | $2{\cdot}05 \times 10^{5}$ | .... |
| Glass | $2{\cdot}60 \times 10^{3}$ | 1400 | .... | $6{\cdot}70 \times 10^{2}$ | .... | .... |
| Ice | $9{\cdot}20 \times 10^{2}$ | 273 | .... | $2{\cdot}10 \times 10^{3}$ | $3{\cdot}34 \times 10^{5}$ | .... |
| Glycerol | $1{\cdot}26 \times 10^{3}$ | 291 | 563 | $2{\cdot}43 \times 10^{3}$ | $1{\cdot}81 \times 10^{5}$ | $8{\cdot}30 \times 10^{5}$ |
| Methanol | $7{\cdot}91 \times 10^{2}$ | 175 | 338 | $2{\cdot}52 \times 10^{3}$ | $9{\cdot}9 \times 10^{4}$ | $1{\cdot}12 \times 10^{6}$ |
| Sea Water | $1{\cdot}02 \times 10^{3}$ | 264 | 377 | $3{\cdot}93 \times 10^{3}$ | .... | .... |
| Water | $1{\cdot}00 \times 10^{3}$ | 273 | 373 | $4{\cdot}19 \times 10^{3}$ | $3{\cdot}34 \times 10^{5}$ | $2{\cdot}26 \times 10^{6}$ |
| Air | 1·29 | .... | .... | .... | .... | .... |
| Hydrogen | $9{\cdot}0 \times 10^{-2}$ | 14 | 20 | $1{\cdot}43 \times 10^{4}$ | .... | $4{\cdot}50 \times 10^{5}$ |
| Nitrogen | 1·25 | 63 | 77 | $1{\cdot}04 \times 10^{3}$ | .... | $2{\cdot}00 \times 10^{5}$ |
| Oxygen | 1·43 | 55 | 90 | $9{\cdot}18 \times 10^{2}$ | .... | $2{\cdot}40 \times 10^{4}$ |

The gas densities refer to a temperature of 273 K and a pressure of $1{\cdot}01 \times 10^{5}$ Pa.

MARKS | DO NOT WRITE IN THIS MARGIN

**Total marks — 140 marks**

**Attempt ALL questions**

1.

Calvin Chan/shutterstock.com

A car on a long straight track accelerates from rest. The car's run begins at time $t = 0$.

Its velocity $v$ at time $t$ is given by the equation

$$v = 0{\cdot}135t^2 + 1{\cdot}26t$$

where $v$ is measured in $\text{m s}^{-1}$ and $t$ is measured in s.

Using **calculus** methods:

(a) determine the acceleration of the car at $t = 15{\cdot}0\,\text{s}$; **3**

*Space for working and answer*

(b) determine the displacement of the car from its original position at this time. **3**

*Space for working and answer*

MARKS | DO NOT WRITE IN THIS MARGIN

**2.** (a) An ideal conical pendulum consists of a mass moving with constant speed in a circular path, as shown in Figure 2A.

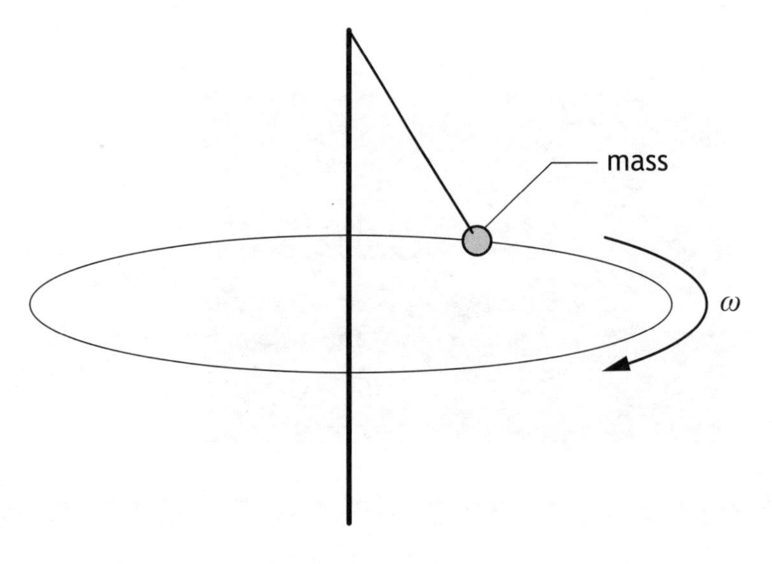


Figure 2A

(i) Explain why the mass is accelerating despite moving with constant speed. **1**

(ii) State the direction of this acceleration. **1**

MARKS | DO NOT WRITE IN THIS MARGIN

**2. (continued)**

(b) Swingball is a garden game in which a ball is attached to a light string connected to a vertical pole as shown in Figure 2B.

The motion of the ball can be modelled as a conical pendulum.

The ball has a mass of 0·059 kg.

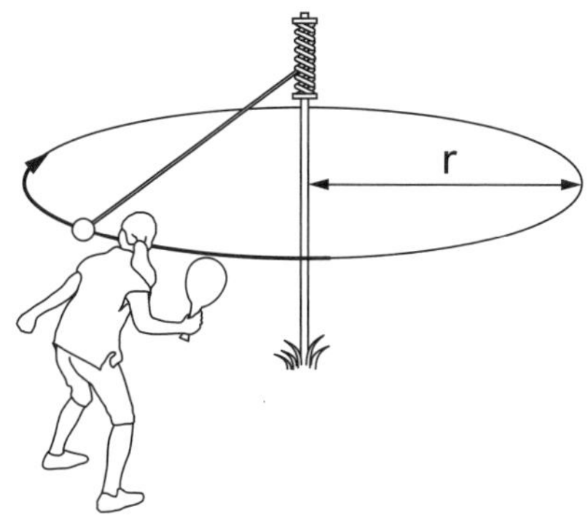


Figure 2B

(i) The ball is hit such that it moves with constant speed in a horizontal circle of radius 0·48 m.

The ball completes 1·5 revolutions in 2·69 s.

(A) Show that the angular velocity of the ball is 3·5 rad $s^{-1}$. **2**

*Space for working and answer*

(B) Calculate the magnitude of the centripetal force acting on the ball. **3**

*Space for working and answer*

MARKS | DO NOT WRITE IN THIS MARGIN

**2. (b) (i) (continued)**

(C) The horizontal component of the tension in the string provides this centripetal force and the vertical component balances the weight of the ball.

Calculate the magnitude of the tension in the string. **3**

*Space for working and answer*

(ii) The string breaks whilst the ball is at the position shown in Figure 2C.

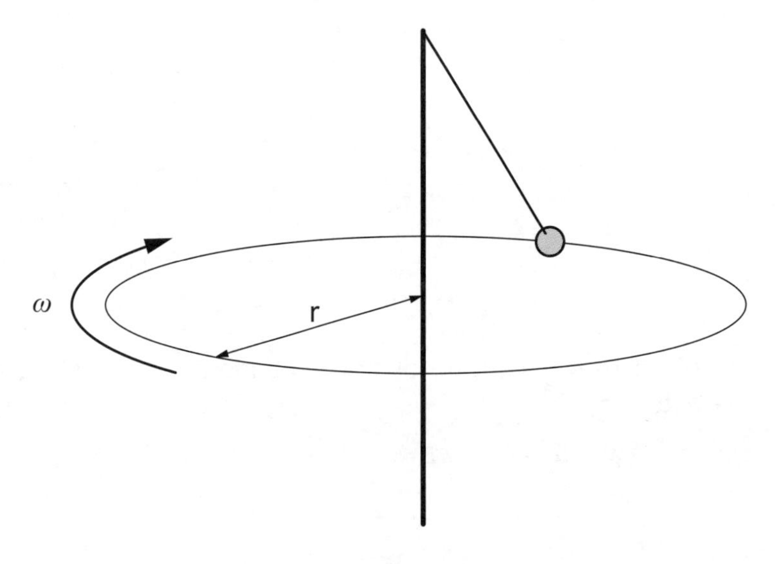


Figure 2C

On Figure 2C, draw the direction of the ball's velocity **immediately** after the string breaks. **1**

(An additional diagram, if required, can be found on *Page thirty-nine*.)

MARKS DO NOT WRITE IN THIS MARGIN

**3.** A spacecraft is orbiting a comet as shown in Figure 3.

The comet can be considered as a sphere with a radius of $2{\cdot}1 \times 10^3$ m and a mass of $9{\cdot}5 \times 10^{12}$ kg.

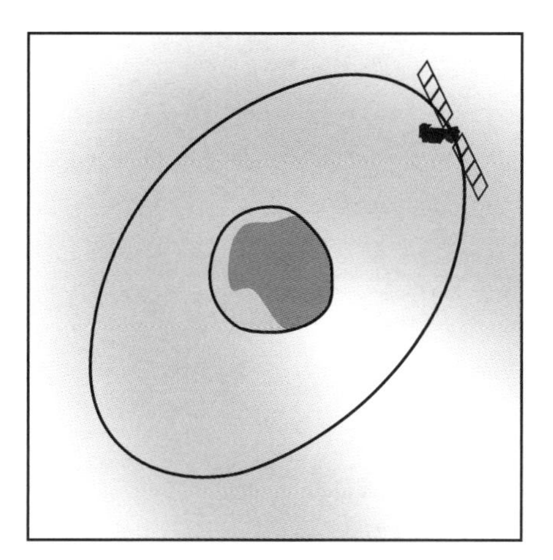

Figure 3 (not to scale)

(a) A lander was released by the spacecraft to land on the surface of the comet. After impact with the comet, the lander bounced back from the surface with an initial upward vertical velocity of $0{\cdot}38\,\mathrm{m\,s^{-1}}$.

By calculating the escape velocity of the comet, show that the lander returned to the surface for a second time. **4**

*Space for working and answer*

[Turn over

MARKS | DO NOT WRITE IN THIS MARGIN

3. **(continued)**

(b) (i) Show that the gravitational field strength at the surface of the comet is $1{\cdot}4 \times 10^{-4}\,\mathrm{N\,kg^{-1}}$. **3**

*Space for working and answer*

(ii) Using the data from the space mission, a student tries to calculate the maximum height reached by the lander after its first bounce.

The student's working is shown below

$$v^2 = u^2 + 2as$$

$$0 = 0{\cdot}38^2 + 2 \times \left(-1{\cdot}4 \times 10^{-4}\right) \times s$$

$$s = 515{\cdot}7\,\mathrm{m}$$

The actual maximum height reached by the lander was **not** as calculated by the student.

State whether the actual maximum height reached would be greater or smaller than calculated by the student.

You must justify your answer. **3**

MARKS DO NOT WRITE IN THIS MARGIN

**4.** Epsilon Eridani is a star $9{\cdot}94 \times 10^{16}$ m from Earth. It has a diameter of $1{\cdot}02 \times 10^{9}$ m. The apparent brightness of Epsilon Eridani is measured on Earth to be $1{\cdot}05 \times 10^{-9}\,\mathrm{W\,m^{-2}}$.

(a) Calculate the luminosity of Epsilon Eridani. **3**

*Space for working and answer*

(b) Calculate the surface temperature of Epsilon Eridani. **3**

*Space for working and answer*

(c) State an assumption made in your calculation in (b). **1**

MARKS | DO NOT WRITE IN THIS MARGIN

**5.** Einstein's theory of general relativity can be used to describe the motion of objects in non-inertial frames of reference. The equivalence principle is a key assumption of general relativity.

(a) Explain what is meant by the terms:

(i) *non-inertial frames of reference*; **1**

(ii) *the equivalence principle*. **1**

(b) Two astronauts are on board a spacecraft in deep space far away from any large masses. When the spacecraft is accelerating one astronaut throws a ball towards the other.

(i) On Figure 5A sketch the path that the ball would follow in the astronauts' frame of reference. **1**

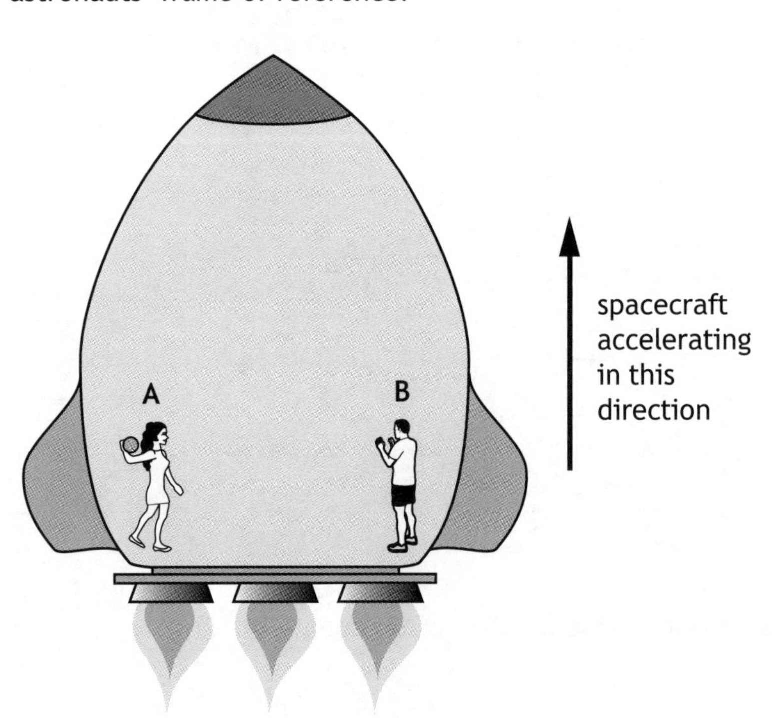


Figure 5A

(An additional diagram, if required, can be found on *Page thirty-nine*.)

MARKS | DO NOT WRITE IN THIS MARGIN

**5. (b) (continued)**

(ii) The experiment is repeated when the spacecraft is travelling at constant speed.

On Figure 5B sketch the path that the ball would follow in the astronauts' frame of reference. **1**

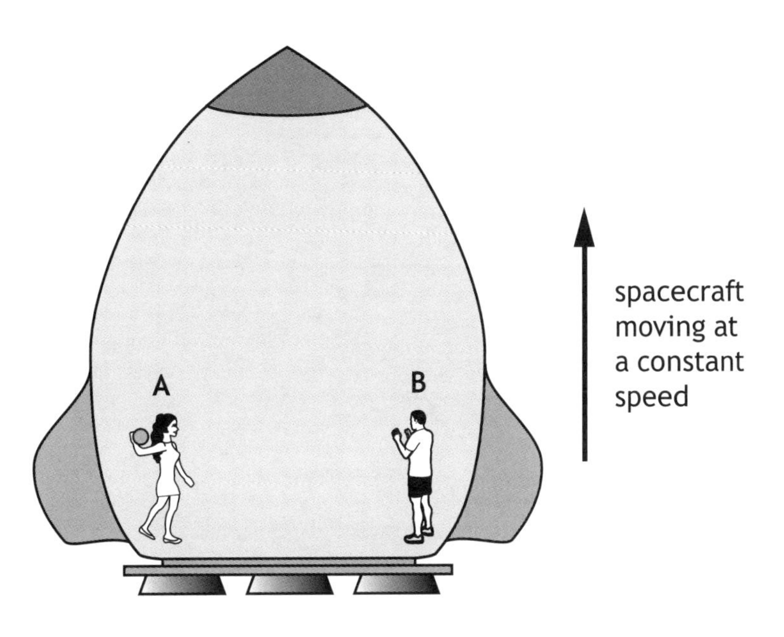


Figure 5B

(An additional diagram, if required, can be found on *Page 40*.)

(c) A clock is on the surface of the Earth and an identical clock is on board a spacecraft which is accelerating in deep space at $8\,\text{m}\,\text{s}^{-2}$.

State which clock runs slower.

Justify your answer in terms of the equivalence principle. **2**

**[Turn over**

MARKS DO NOT WRITE IN THIS MARGIN

6. A student makes the following statement.

"Quantum theory — I don't understand it. I don't really know what it is. I believe that classical physics can explain everything."

Use your knowledge of physics to comment on the statement. **3**

MARKS | DO NOT WRITE IN THIS MARGIN

**7.** (a) The Earth can be modelled as a black body radiator.

The average surface temperature of the Earth can be estimated using the relationship

$$T = \frac{b}{\lambda_{\text{peak}}}$$

where

$T$ is the average surface temperature of the Earth in kelvin;

$b$ is Wien's Displacement Constant equal to $2{\cdot}89 \times 10^{-3}$ K m;

$\lambda_{\text{peak}}$ is the peak wavelength of the radiation emitted by a black body radiator.

The average surface temperature of Earth is 15 °C.

(i) Estimate the peak wavelength of the radiation emitted by Earth. **3**

*Space for working and answer*

(ii) To which part of the electromagnetic spectrum does this peak wavelength correspond? **1**

**[Turn over**

MARKS DO NOT WRITE IN THIS MARGIN

**7. (continued)**

(b) In order to investigate the properties of black body radiators a student makes measurements from the spectra produced by a filament lamp. Measurements are made when the lamp is operated at its rated voltage and when it is operated at a lower voltage.

The filament lamp can be considered to be a black body radiator.

A graph of the results obtained is shown in Figure 7.

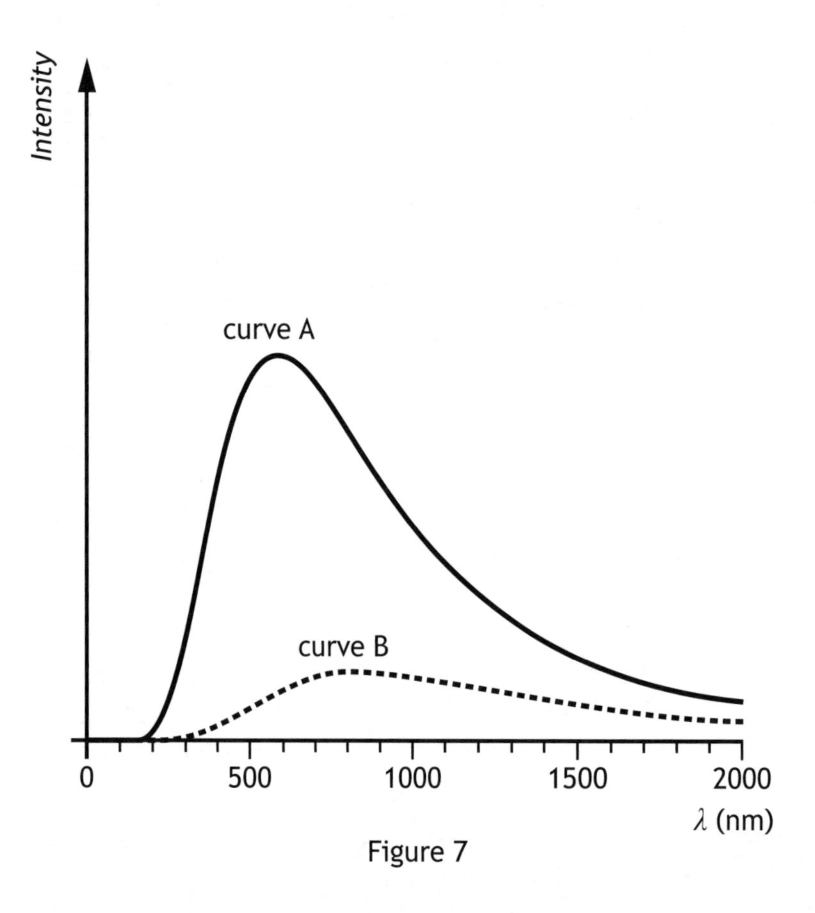


Figure 7

(i) State which curve corresponds to the radiation emitted when the filament lamp is operating at its rated voltage.

You must justify your answer. **2**

(ii) The shape of the curves on the graph on Figure 7 is not as predicted by classical physics.

On Figure 7, sketch a curve to show the result predicted by classical physics. **1**

(An additional graph, if required, can be found on *Page forty*.)

MARKS | DO NOT WRITE IN THIS MARGIN

**8.** Werner Heisenberg is considered to be one of the pioneers of quantum mechanics.

He is most famous for his uncertainty principle which can be expressed in the equation

$$\Delta x \Delta p_x \geq \frac{h}{4\pi}$$

(a) (i) State what quantity is represented by the term $\Delta p_x$. **1**

(ii) Explain the implications of the Heisenberg uncertainty principle for experimental measurements. **1**

[Turn over

MARKS | DO NOT WRITE IN THIS MARGIN

**8. (continued)**

(b) In an experiment to investigate the nature of particles, individual electrons were fired one at a time from an electron gun through a narrow double slit. The position where each electron struck the detector was recorded and displayed on a computer screen.

The experiment continued until a clear pattern emerged on the screen as shown in Figure 8.

The momentum of each electron at the double slit is $6{\cdot}5 \times 10^{-24}$ kg m s$^{-1}$.

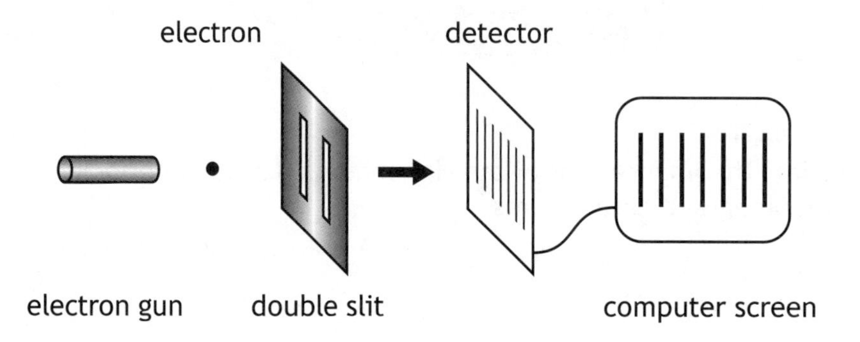


Figure 8 not to scale

(i) The experimenter had three different double slits with slit separations 0·1 mm, 0·1 μm and 0·1 nm.

State which double slit was used to produce the image on the screen.

You must justify your answer by calculation of the de Broglie wavelength. **4**

*Space for working and answer*

MARKS | DO NOT WRITE IN THIS MARGIN

**8. (b) (continued)**

(ii) The uncertainty in the momentum of an electron at the double slit is $6{\cdot}5 \times 10^{-26}\,\text{kg m s}^{-1}$.

Calculate the minimum absolute uncertainty in the position of the electron. **3**

*Space for working and answer*

(iii) Explain fully how the experimental result shown in Figure 8 can be interpreted. **3**

**[Turn over**

MARKS | DO NOT WRITE IN THIS MARGIN

**9.** A particle with charge $q$ and mass $m$ is travelling with constant speed $v$. The particle enters a uniform magnetic field at 90° and is forced to move in a circle of radius $r$ as shown in Figure 9.

The magnetic induction of the field is $B$.

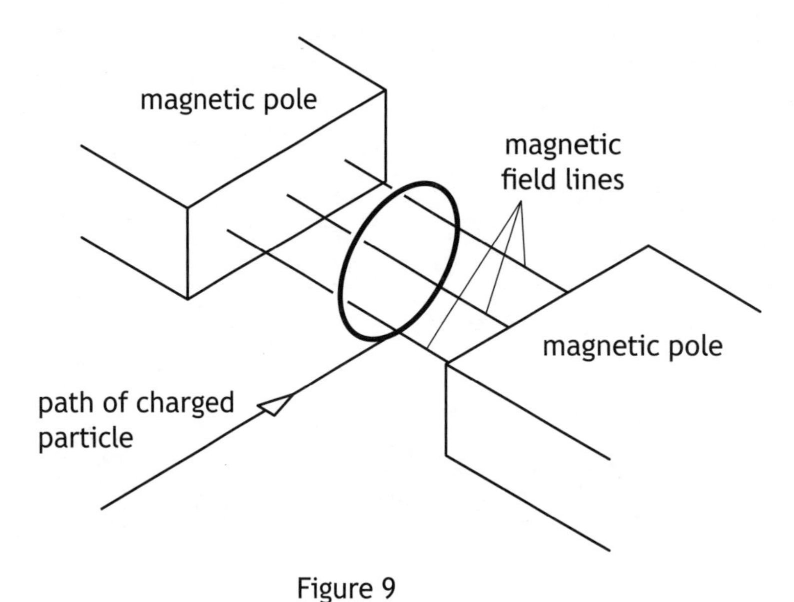


Figure 9

(a) Show that the radius of the circular path of the particle is given by

$$r = \frac{mv}{Bq}$$

2

MARKS DO NOT WRITE IN THIS MARGIN

**9. (continued)**

(b) In an experimental nuclear reactor, charged particles are contained in a magnetic field. One such particle is a deuteron consisting of one proton and one neutron.

The kinetic energy of each deuteron is 1·50 MeV.

The mass of the deuteron is $3{\cdot}34 \times 10^{-27}$ kg.

Relativistic effects can be ignored.

(i) Calculate the speed of the deuteron. **4**

*Space for working and answer*

(ii) Calculate the magnetic induction required to keep the deuteron moving in a circular path of radius 2·50 m. **2**

*Space for working and answer*

**[Turn over**

MARKS DO NOT WRITE IN THIS MARGIN

**9. (b) (continued)**

(iii) Deuterons are fused together in the reactor to produce isotopes of helium.

$^{3}_{2}He$ nuclei, each comprising 2 protons and 1 neutron, are present in the reactor.

A $^{3}_{2}He$ nucleus also moves in a circular path in the same magnetic field.

The $^{3}_{2}He$ nucleus moves at the same speed as the deuteron.

State whether the radius of the circular path of the $^{3}_{2}He$ nucleus is greater than, equal to or less than 2·50 m.

You must justify your answer. **2**

MARKS | DO NOT WRITE IN THIS MARGIN

**10.** (a) (i) State what is meant by *simple harmonic motion*. **1**

(ii) The displacement of an oscillating object can be described by the expression

$$y = A\cos\omega t$$

where the symbols have their usual meaning.

Show that this expression is a solution to the equation

$$\frac{d^2y}{dt^2} + \omega^2 y = 0$$

**2**

**[Turn over**

MARKS | DO NOT WRITE IN THIS MARGIN

**10. (continued)**

(b) A mass of 1·5 kg is suspended from a spring of negligible mass as shown in Figure 10. The mass is displaced downwards 0·040 m from its equilibrium position.

The mass is then released from this position and begins to oscillate. The mass completes ten oscillations in a time of 12 s.

Frictional forces can be considered to be negligible.

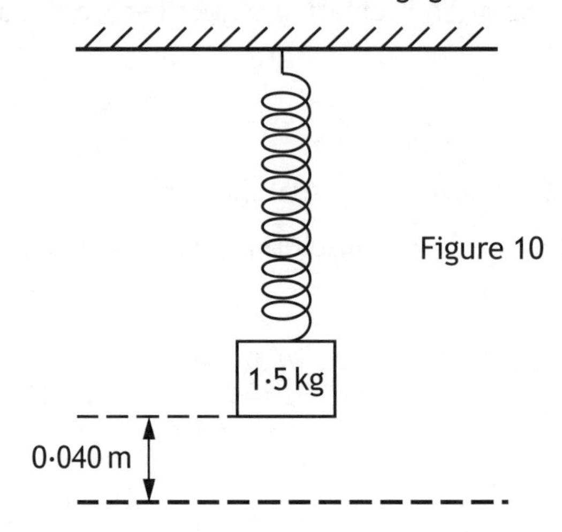


Figure 10

(i) Show that the angular frequency $\omega$ of the mass is $5 \cdot 2\ \text{rad s}^{-1}$. **3**

*Space for working and answer*

(ii) Calculate the maximum velocity of the mass. **3**

*Space for working and answer*

MARKS | DO NOT WRITE IN THIS MARGIN

**10. (b) (continued)**

(iii) Determine the potential energy stored in the spring when the mass is at its maximum displacement. **3**

*Space for working and answer*

(c) The system is now modified so that a damping force acts on the oscillating mass.

(i) Describe how this modification may be achieved. **1**

(ii) Using the axes below sketch a graph showing, for the modified system, how the displacement of the mass varies with time after release.

Numerical values are **not** required on the axes. **1**

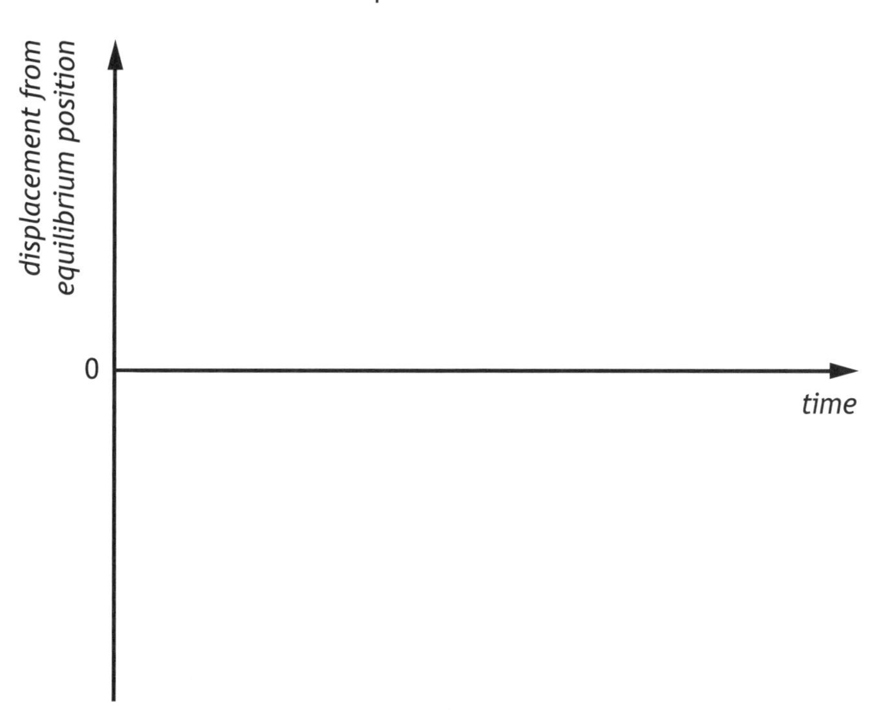


(An additional graph, if required, can be found on *Page forty-one*.)

MARKS DO NOT WRITE IN THIS MARGIN

**11.**

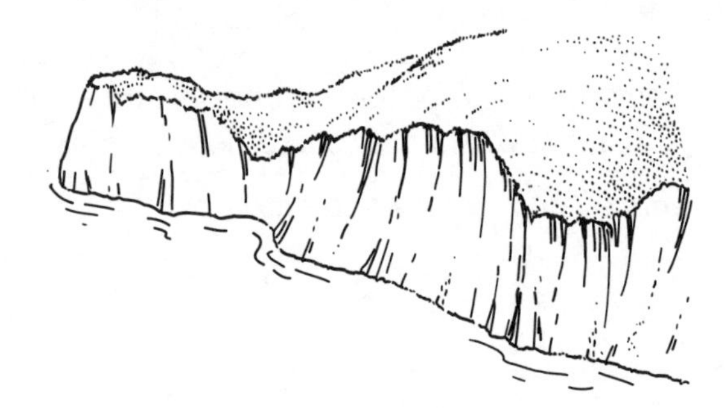

A ship emits a blast of sound from its foghorn. The sound wave is described by the equation

$$y = 0{\cdot}250 \sin 2\pi(118t - 0{\cdot}357x)$$

where the symbols have their usual meaning.

(a) Determine the speed of the sound wave. **4**

*Space for working and answer*

MARKS DO NOT WRITE IN THIS MARGIN

**11. (continued)**

(b) The sound from the ship's foghorn reflects from a cliff. When it reaches the ship this reflected sound has half the energy of the original sound.

Write an equation describing the reflected sound wave at this point. 4

[Turn over

MARKS | DO NOT WRITE IN THIS MARGIN

12. Some early 3D video cameras recorded two separate images at the same time to create two almost identical movies.

Cinemas showed 3D films by projecting these two images simultaneously onto the same screen using two projectors. Each projector had a polarising filter through which the light passed as shown in Figure 12.

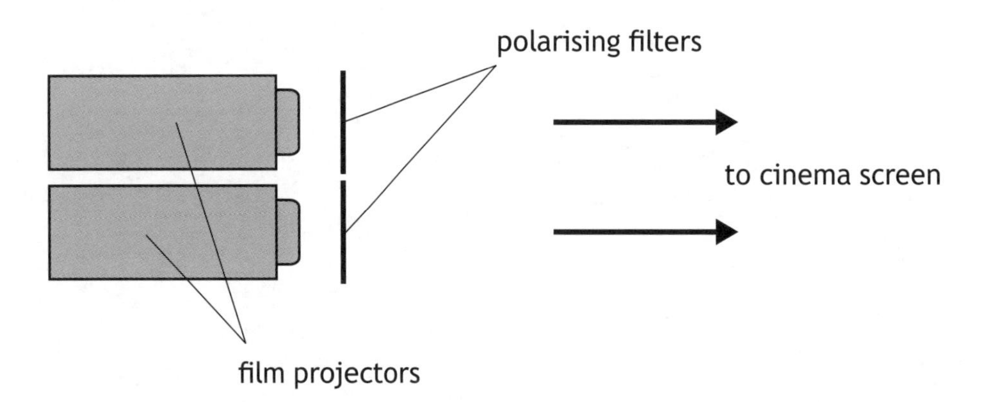


Figure 12

(a) Describe how the transmission axes of the two polarising filters should be arranged so that the two images on the screen do not interfere with each other. **1**

(b) A student watches a 3D movie using a pair of glasses which contains two polarising filters, one for each eye.

Explain how this arrangement enables a different image to be seen by each eye. **2**

MARKS | DO NOT WRITE IN THIS MARGIN

**12. (continued)**

(c) Before the film starts, the student looks at a ceiling lamp through one of the filters in the glasses. While looking at the lamp, the student then rotates the filter through 90°.

State what effect, if any, this rotation will have on the observed brightness of the lamp.

Justify your answer. **2**

(d) During the film, the student looks at the screen through only one of the filters in the glasses. The student then rotates the filter through 90° and does not observe any change in brightness.

Explain this observation. **1**

**[Turn over**

MARKS | DO NOT WRITE IN THIS MARGIN

**13.** (a) $Q_1$ is a point charge of +12 nC. Point Y is 0·30 m from $Q_1$ as shown in Figure 13A.

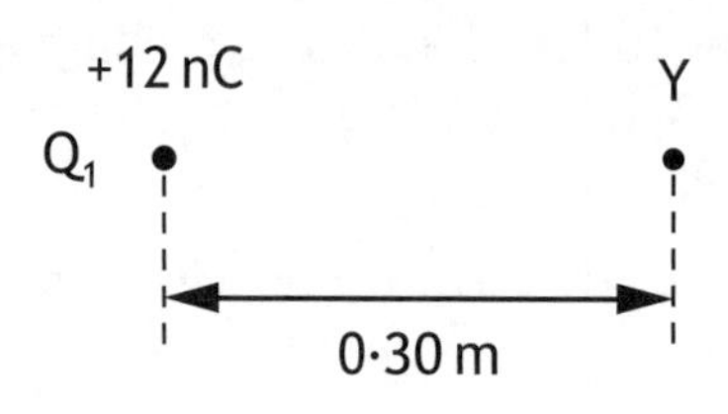


Figure 13A

Show that the electrical potential at point Y is +360 V. **2**

*Space for working and answer*

(b) A second point charge $Q_2$ is placed at a distance of 0·40 m from point Y as shown in Figure 13B. The electrical potential at point Y is now zero.

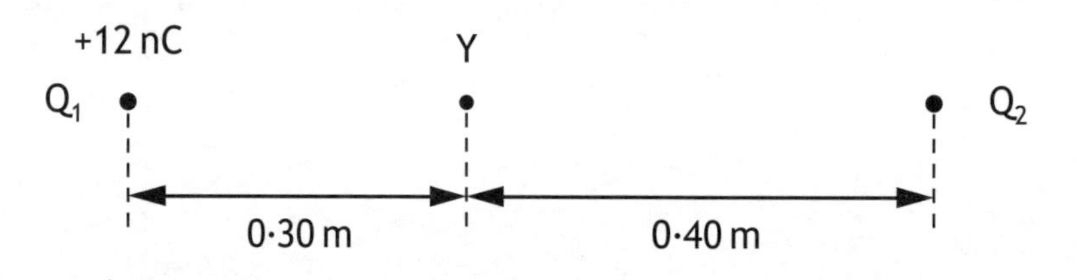


Figure 13B

(i) Determine the charge of $Q_2$. **3**

*Space for working and answer*

MARKS | DO NOT WRITE IN THIS MARGIN

**13. (b) (continued)**

(ii) Determine the electric field strength at point Y. **4**

*Space for working and answer*

(iii) On Figure 13C, sketch the electric field pattern for this system of charges. **2**

$Q_1$ • • $Q_2$

Figure 13C

(An additional diagram, if required, can be found on *Page forty-one*.)

**[Turn over**

MARKS | DO NOT WRITE IN THIS MARGIN

**14.** A student measures the magnetic induction at a distance $r$ from a long straight current carrying wire using the apparatus shown in Figure 14.

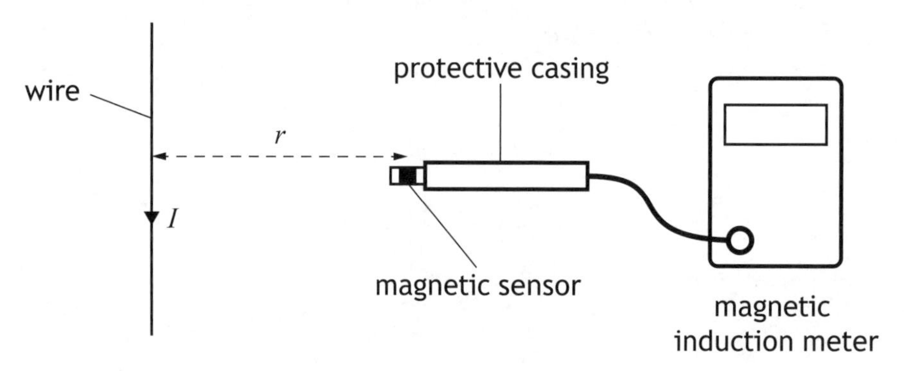


Figure 14

The following data are obtained.

Distance from wire $r = 0{\cdot}10\,\text{m}$
Magnetic induction $B = 5{\cdot}0\,\mu\text{T}$

(a) Use the data to calculate the current $I$ in the wire. **3**

*Space for working and answer*

(b) The student estimates the following uncertainties in the measurements of $B$ and $r$.

| Uncertainties in $r$ | | Uncertainties in $B$ | |
|---|---|---|---|
| reading | $\pm 0{\cdot}002\,\text{m}$ | reading | $\pm 0{\cdot}1\,\mu\text{T}$ |
| calibration | $\pm 0{\cdot}0005\,\text{m}$ | calibration | $\pm 1{\cdot}5\%$ of reading |

(i) Calculate the percentage uncertainty in the measurement of $r$. **1**

*Space for working and answer*

MARKS DO NOT WRITE IN THIS MARGIN

**14. (b) (continued)**

(ii) Calculate the percentage uncertainty in the measurement of $B$. **3**

*Space for working and answer*

(iii) Calculate the absolute uncertainty in the value of the current in the wire. **2**

*Space for working and answer*

(c) The student measures distance $r$, as shown in Figure 14, using a metre stick. The smallest scale division on the metre stick is 1 mm.

Suggest a reason why the student's estimate of the reading uncertainty in $r$ is not $\pm 0{\cdot}5$ mm. **1**

**[Turn over**

**15.** A student constructs a simple air-insulated capacitor using two parallel metal plates, each of area A, separated by a distance $d$. The plates are separated using small insulating spacers as shown in Figure 15A.

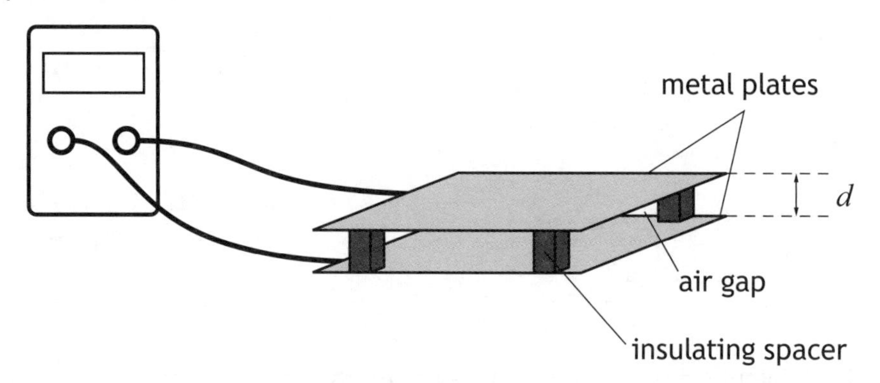


Figure 15A

The capacitance $C$ of the capacitor is given by

$$C = \varepsilon_0 \frac{A}{d}$$

The student investigates how the capacitance depends on the separation of the plates. The student uses a capacitance meter to measure the capacitance for different plate separations. The plate separation is measured using a ruler.

The results are used to plot the graph shown in Figure 15B.

The area of each metal plate is $9{\cdot}0 \times 10^{-2}\,\text{m}^2$.

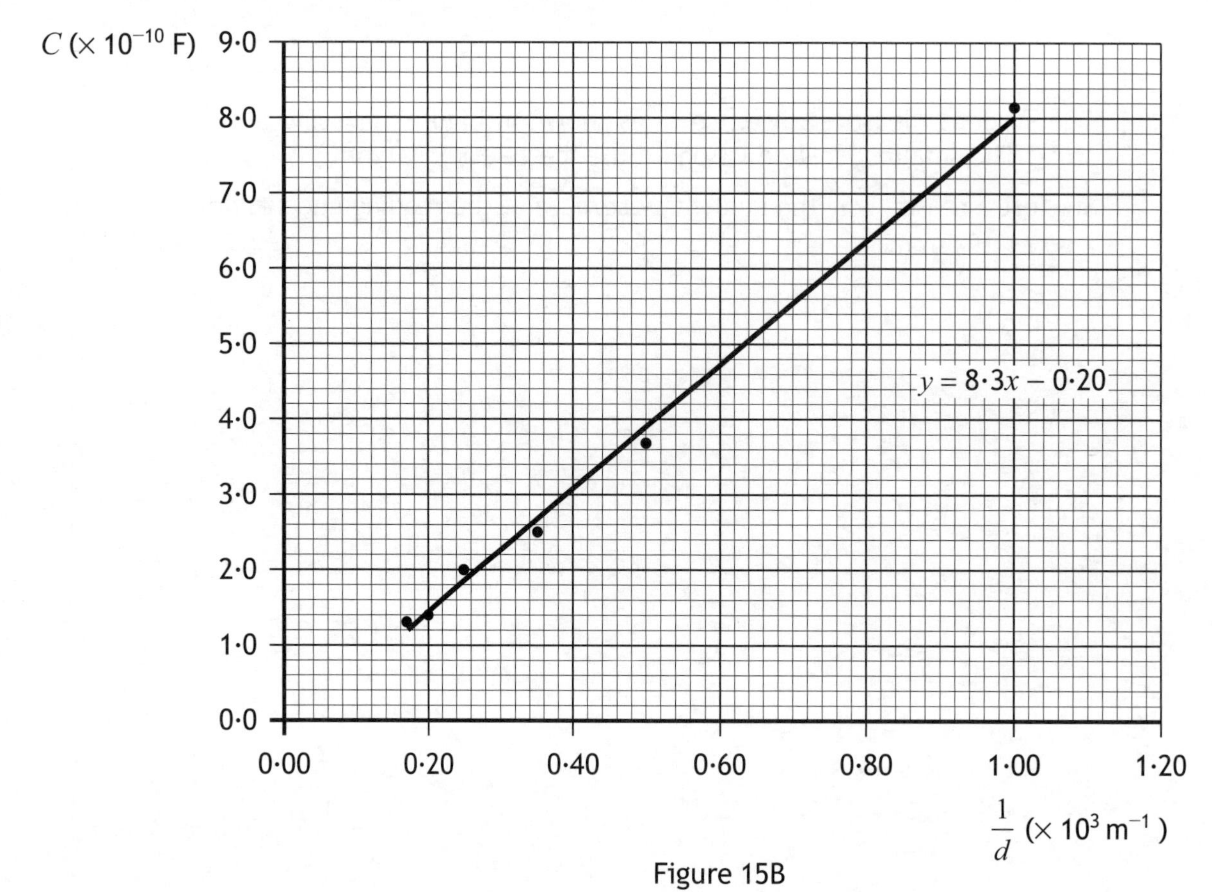


Figure 15B

MARKS DO NOT WRITE IN THIS MARGIN

**15. (continued)**

(a) (i) Use information from the graph to determine a value for $\varepsilon_0$, the permittivity of free space. **3**

*Space for working and answer*

(ii) Use your calculated value for the permittivity of free space to determine a value for the speed of light in air. **3**

*Space for working and answer*

(b) The best fit line on the graph does not pass through the origin as theory predicts.

Suggest a reason for this. **1**

**[Turn over**

[BLANK PAGE]

DO NOT WRITE ON THIS PAGE

MARKS DO NOT WRITE IN THIS MARGIN

**16.** A student uses two methods to determine the moment of inertia of a solid sphere about an axis through its centre.

(a) In the first method the student measures the mass of the sphere to be 3·8 kg and the radius to be 0·053 m.

Calculate the moment of inertia of the sphere. **3**

*Space for working and answer*

(b) In the second method, the student uses conservation of energy to determine the moment of inertia of the sphere.

The following equation describes the conservation of energy as the sphere rolls down the slope

$$mgh = \frac{1}{2}mv^2 + \frac{1}{2}I\omega^2$$

where the symbols have their usual meanings.

The equation can be rearranged to give the following expression

$$2gh = \left(\frac{I}{mr^2} + 1\right)v^2$$

This expression is in the form of the equation of a straight line through the origin,

$$y = gradient \times x$$

[Turn over

MARKS DO NOT WRITE IN THIS MARGIN

**16. (b) (continued)**

The student measures the height of the slope $h$. The student then allows the sphere to roll down the slope and measures the final speed of the sphere $v$ at the bottom of the slope as shown in Figure 16.

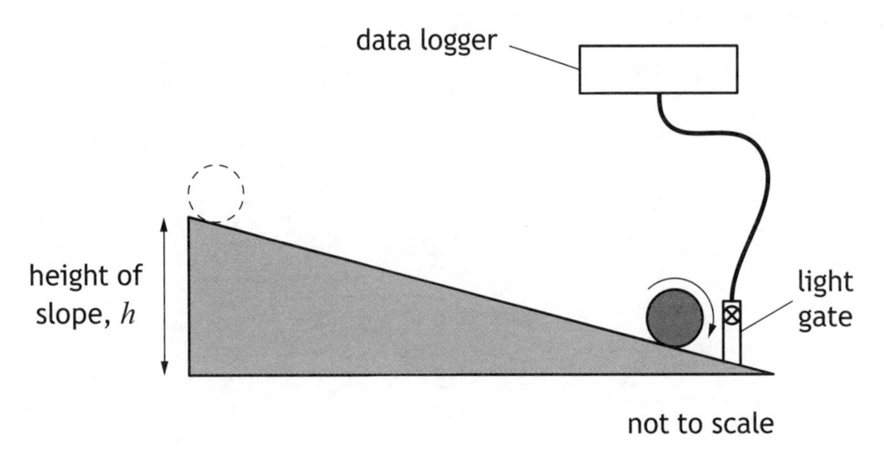


Figure 16

The following is an extract from the student's notebook.

| $h$ (m) | $v$ (m s$^{-1}$) | $2gh$ (m$^2$ s$^{-2}$) | $v^2$ (m$^2$ s$^{-2}$) |
|---|---|---|---|
| 0·020 | 0·42 | 0·39 | 0·18 |
| 0·040 | 0·63 | 0·78 | 0·40 |
| 0·060 | 0·68 | 1·18 | 0·46 |
| 0·080 | 0·95 | 1·57 | 0·90 |
| 0·100 | 1·05 | 1·96 | 1·10 |

$m = 3{\cdot}8$ kg $\quad r = 0{\cdot}053$ m

(i) On the square-ruled paper on *Page thirty-seven*, draw a graph that would allow the student to determine the moment of inertia of the sphere. **3**

(ii) Use the gradient of your line to determine the moment of inertia of the sphere. **3**

*Space for working and answer*

(An additional square-ruled paper, if required, can be found on *Page forty-two*.)

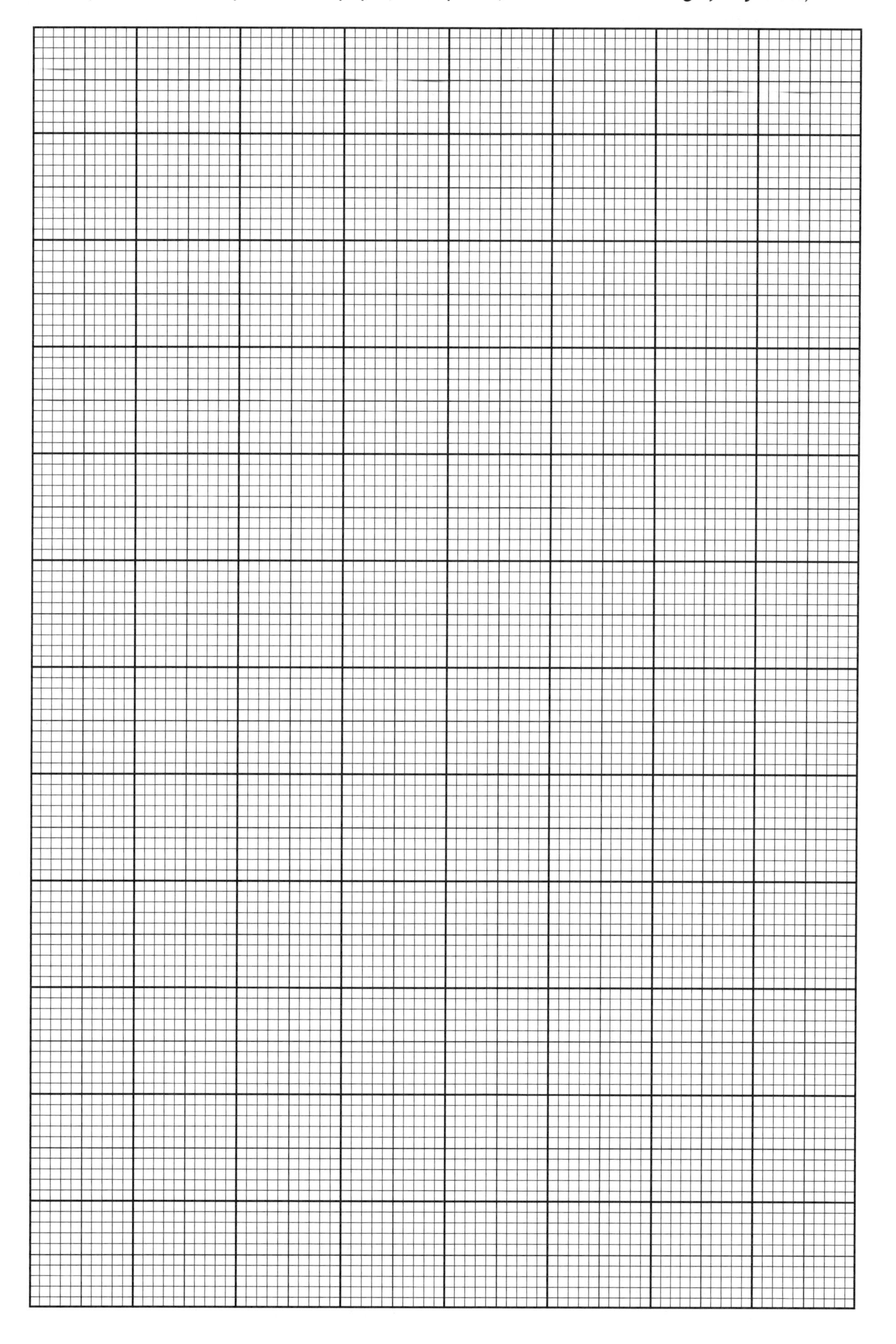

**[Turn over for next question**

MARKS DO NOT WRITE IN THIS MARGIN

**16. (continued)**

(c) The student states that more confidence should be placed in the value obtained for the moment of inertia in the second method.

Use your knowledge of experimental physics to comment on the student's statement. **3**

**[END OF QUESTION PAPER]**

MARKS DO NOT WRITE IN THIS MARGIN

## ADDITIONAL SPACE FOR ANSWERS AND ROUGH WORK

Additional diagram for Question 2 (b) (ii)

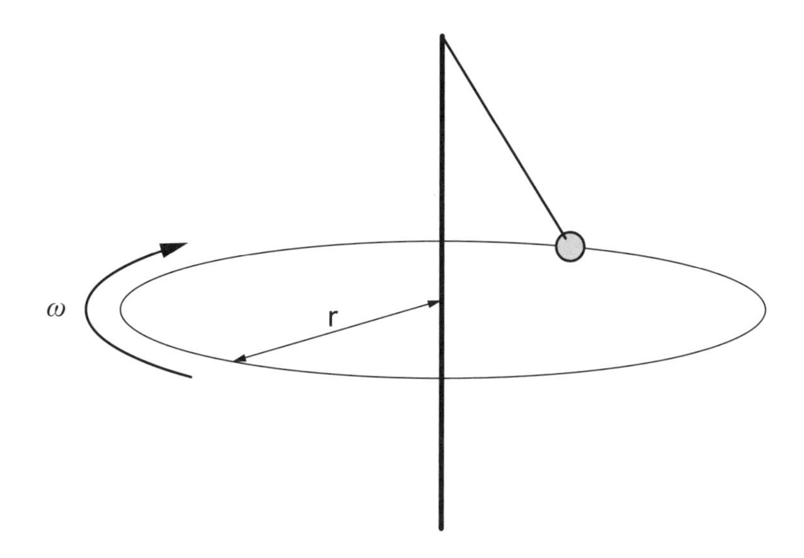


Figure 2C

Additional diagram for Question 5 (b) (i)

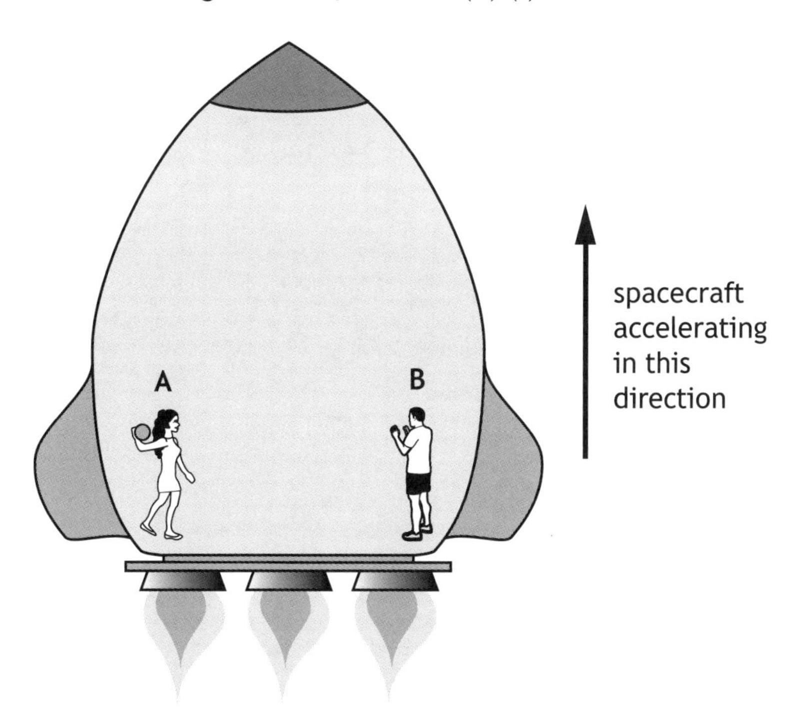


Figure 5A

MARKS | DO NOT WRITE IN THIS MARGIN

**ADDITIONAL SPACE FOR ANSWERS AND ROUGH WORK**

Additional diagram for Question 5 (b) (ii)

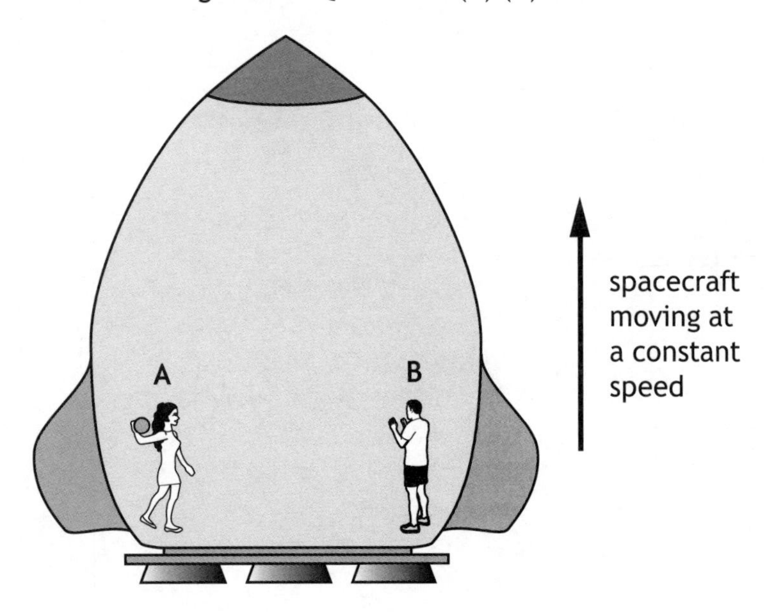


Figure 5B

Additional diagram for Question 7 (b) (ii)

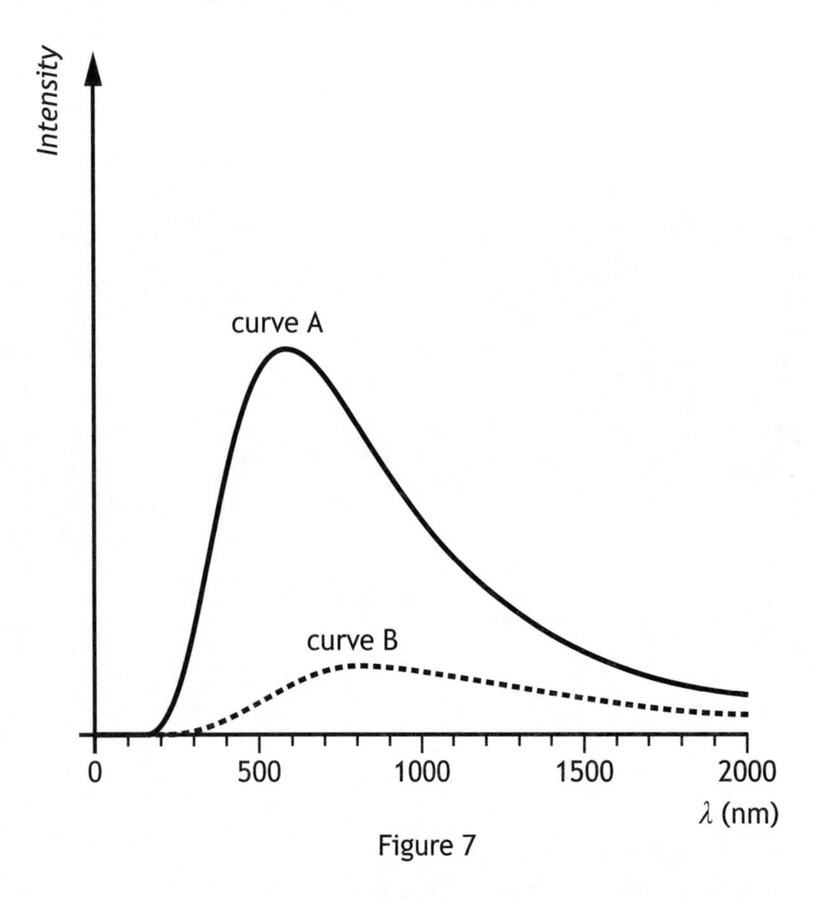


Figure 7

MARKS | DO NOT WRITE IN THIS MARGIN

## ADDITIONAL SPACE FOR ANSWERS AND ROUGH WORK

Additional diagram for Question 10 (c) (ii)

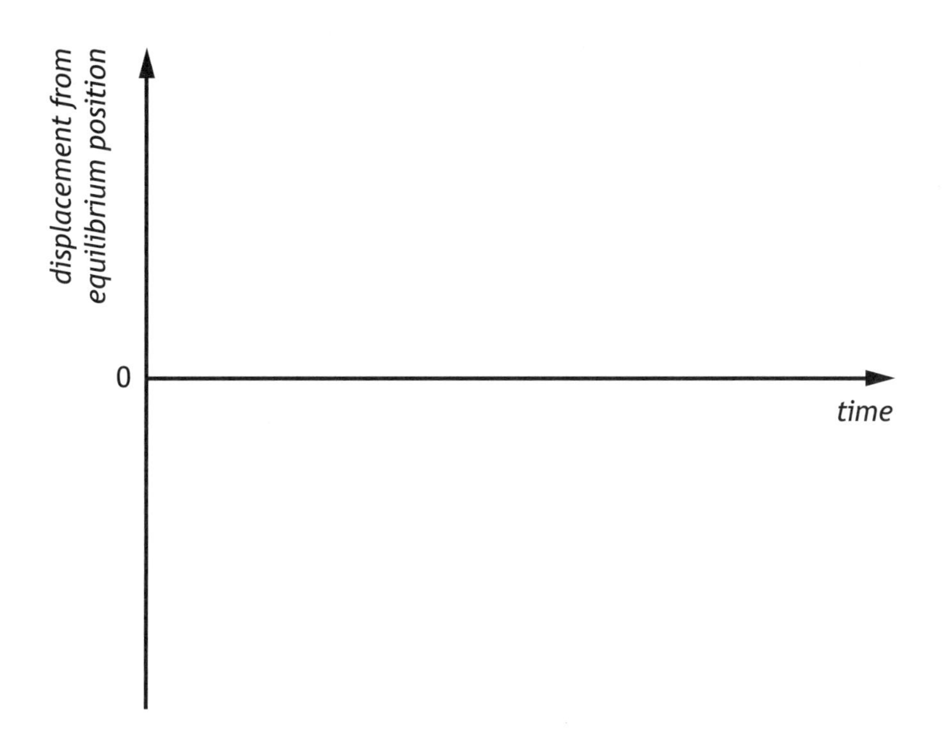


Additional diagram for Question 13 (b) (iii)

$Q_1$ • • $Q_2$

Figure 13C

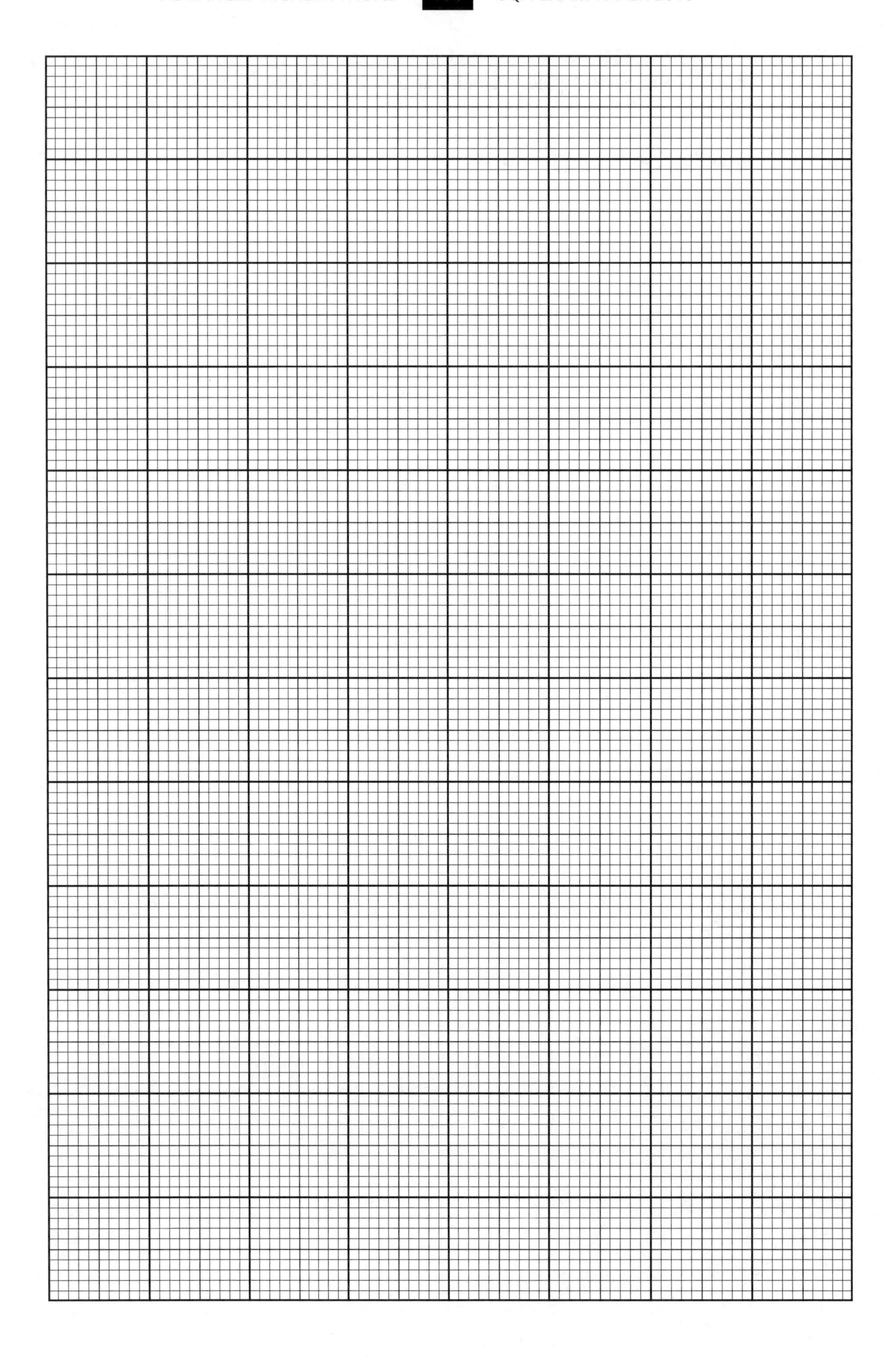

MARKS DO NOT WRITE IN THIS MARGIN

**ADDITIONAL SPACE FOR ANSWERS AND ROUGH WORK**

MARKS DO NOT WRITE IN THIS MARGIN

**ADDITIONAL SPACE FOR ANSWERS AND ROUGH WORK**

ADVANCED HIGHER

# Answers

# ANSWERS FOR

# SQA & HODDER GIBSON ADVANCED HIGHER PHYSICS 2016

## ADVANCED HIGHER PHYSICS 2015 SPECIMEN QUESTION PAPER

| Question | | | Expected response | Max mark | Additional guidance |
|---|---|---|---|---|---|
| 1. | (a) | | $\omega = \frac{\theta}{t}$ (1)<br>$= \frac{1250 \times 2 \times \pi}{60}$ (1)<br>$= 131\ \text{rad s}^{-1}$ | 2 | If final answer is not shown then maximum of 1 mark can be awarded. |
| | (b) | (i) | $\alpha = \frac{\omega_1 - \omega_0}{t}$ (1)<br>$= \frac{131 - 7{\cdot}50}{12}$ (1)<br>$= 10{\cdot}3\ \text{rad s}^{-2}$ (1) | 3 | Accept:<br>10<br>10·3<br>10·29<br>10·292 |
| | | (ii) | $\theta = \omega_0 t + \frac{1}{2}\alpha t^2$ (1)<br>(1)<br>$= 7{\cdot}50 \times 12{\cdot}0 + 0{\cdot}5 \times 10{\cdot}3 \times 12{\cdot}0^2$<br>$= 831{\cdot}6$ (rad) (1)<br>(1)<br>revolutions $= \frac{831{\cdot}6}{2\pi}$<br>$= 132$ (1) | 5 | If candidate stops here unit must be present for mark 3.<br>Accept:<br>130<br>132<br>132·4<br>132·35 |
| | (c) | | centripetal force $= m\omega^2 r$ (1)<br>$= 1{\cdot}5 \times 10^{-2} \times 131^2 \times 0{\cdot}28$ (1)<br>$= 72\ \text{N}$ (1) | 3 | Accept:<br>70<br>72<br>72·1<br>72·08 |
| | (d) | (i) | The drum exerts a centripetal/central force on the clothing. (1)<br>No centripetal/central force acting on water. (1) | 2 | |
| | | (ii) | <br>(1) | 1 | |
| | | (iii) | Centripetal force decreases ... (1)<br>as mass of wet clothing decreases (1) | 2 | |
| 2. | (a) | | $I = \frac{1}{2}mr^2$ (1)<br>$= 0{\cdot}5 \times 6{\cdot}0 \times 0{\cdot}50^2$ (1)<br>$= 0{\cdot}75\ \text{kg m}^2$ | 2 | If final answer is not shown then maximum of 1 mark can be awarded. |

| Question | | | Expected response | Max mark | Additional guidance |
|---|---|---|---|---|---|
| | (b) | (i) | 2 kg mass:<br>$I = mr^2$ (1)<br>$= 2{\cdot}0 \times 0{\cdot}40^2$ (1)<br>$= 0{\cdot}32\ (\text{kgm}^2)$<br>$\text{Total} = 0{\cdot}32 + 0{\cdot}75 = 1{\cdot}1\,\text{kgm}^2$ (1) | 3 | Accept:<br>1<br>1·1<br>1·07<br>1·070 |
| | | (ii) | $I_1\omega_1 = I_2\omega_2$ (1)<br>$0{\cdot}75 \times 12 = 1{\cdot}1 \times \omega_2$ (1)<br>$\omega_2 = 8{\cdot}2\,\text{rad s}^{-1}$ (1) | 3 | Accept:<br>8<br>8·2<br>8·18<br>8·182<br>Also accept 8·4 if 1·07 is clearly used. |
| | | (iii) | No external torque acts on system.<br>**Or**, 2 kg can be considered as a point mass. | 1 | |
| | (c) | | • the (final) angular velocity will be greater (1)<br>• the final moment of inertia is less than in b(ii) (1) | 2 | Reference must be made to moment of inertia for the second mark.<br>Insufficient to say "sphere rolls off" without effect on moment of inertia. |
| 3. | (a) | (i) | $\frac{GM_E m}{r^2} = m\omega^2 r$<br>$\omega = \frac{2\pi}{T}$ (1)<br>$\frac{GM_E m}{r^2} = m\frac{4\pi^2}{T^2}r$ (1)<br>$T = 2\pi\sqrt{\frac{r^3}{GM_g}}$ | 2 | To access any marks candidates must start with equating the forces/acceleration.<br>A maximum of 1 mark if final equation is not shown. |
| | | (ii) | $T = 2\pi\sqrt{\frac{(6{\cdot}4\times10^6 + 4{\cdot}0\times10^5)^3}{6{\cdot}67\times10^{-11}\times6{\cdot}0\times10^{24}}}$ (1)<br>$= 5{\cdot}6 \times 10^3\,\text{s}$ (1) | 2 | Accept:<br>6<br>5·6<br>5·57<br>5·569 |
| | (b) | (i) | Value from graph $4{\cdot}15 \times 10^5$ (m) (1)<br>$mg = \frac{GM_E m}{r^2}$ (1)<br>$g = \frac{GM_E}{r^2}$ (1)<br>$= \frac{6{\cdot}67\times10^{-11}\times6{\cdot}0\times10^{24}}{(4{\cdot}15\times10^5 + 6{\cdot}4\times10^6)^2}$<br>$= 8{\cdot}6\,\text{N kg}^{-1}$ (1) | 4 | Accept:<br>9<br>8·6<br>8·62<br>8·617 |
| | | (ii) | Less atmospheric drag/friction **or** will reduce running costs. (1) | 1 | |

| Question | | | Expected response | Max mark | Additional guidance |
|---|---|---|---|---|---|
| | (c) | | The gravitational field is smaller at the ISS (compared to Earth). **(1)**<br>The clocks on ISS will run faster (than those on Earth). **(1)** | **2** | |
| **4.** | (a) | | Betelgeuse will look red-orange. | **1** | |
| | (b) | (i)(A) | $L = 4\pi r^2 \sigma T^4$ **(1)**<br>$L = 4 \times \pi \times (5{\cdot}49 \times 10^{10})^2 \times 5{\cdot}67 \times 10^{-8} \times (1{\cdot}20 \times 10^4)^4$ **(1)**<br>$L = 4{\cdot}45 \times 10^{31}\,\text{W}$ **(1)** | **3** | Accept:<br>$4{\cdot}5 \times 10^{31}$<br>$4{\cdot}45 \times 10^{31}$<br>$4{\cdot}453 \times 10^{31}$<br>$4{\cdot}4531 \times 10^{31}$ |
| | | (i)(B) | Rigel/stars behave as black body(ies). | **1** | |
| | | (ii) | $\%\Delta r = \frac{0{\cdot}50}{5{\cdot}49} \times 100\% = 9{\cdot}1\%$ **(1)**<br>$\%\Delta T = \frac{0{\cdot}05}{1{\cdot}20} \times 100\% = 4{\cdot}2\%$ **(1)**<br>Total $\%\Delta = \sqrt{(9{\cdot}1 \times 2)^2 + (4{\cdot}2 \times 4)^2}$ **(1)**<br>$= 25\%$<br>$\Delta L = 4{\cdot}45 \times 10^{31} \times 0{\cdot}25 = 1 \times 10^{31}\,\text{W}$ **(1)** | **4** | Accept:<br>1·1<br>1·11 |
| | (c) | | $b = \frac{L}{4\pi r^2}$ **(1)**<br>$= \frac{4{\cdot}45 \times 10^{31}}{4 \times \pi \times (773 \times 365 \times 24 \times 60 \times 60 \times 3{\cdot}00 \times 10^8)^2}$ **(1)**<br>For ly to m conversion **(1)**<br>$= 6{\cdot}62 \times 10^{-8}\,\text{Wm}^{-2}$ **(1)** | **4** | Accept:<br>$6{\cdot}6 \times 10^{-8}$<br>$6{\cdot}62 \times 10^{-8}$<br>$6{\cdot}621 \times 10^{-8}$<br>$6{\cdot}6212 \times 10^{-8}$<br>The use of 3·14 or 365·25 may give 6·61. |
| | (d) | | Any two from:<br>• (Most) hydrogen fusion has stopped.<br>• Radius has (significantly) increased.<br>• Surface temperature has decreased.<br>• Core gets hotter. | **2** | |
| **5.** | (a) | (i) | Acceleration is proportional to displacement (from a fixed point) and is always directed to (that) fixed point.<br>**OR**<br>The unbalanced force is proportional to the displacement (from a fixed point) and is always directed to (that) fixed point. | **1** | Accept:<br>$F = -kx$<br>or<br>$a = -kx$ |
| | | (ii) | From graph $T = 5{\cdot}0$ (s) **(1)**<br>$\rightarrow f = \frac{1}{T} = \frac{1}{5{\cdot}0} = 0{\cdot}20\,\text{s}$<br>$\omega = 2\pi f$ **(1)**<br>$= 2 \times \pi \times 0{\cdot}20$ **(1)**<br>$= 1{\cdot}3\,\text{rad}\,\text{s}^{-1}$ **(1)** | **4** | Accept:<br>1<br>1·3<br>1·26<br>1·257<br>Use of $\omega = \frac{2\pi}{T}$ is possible. |
| | | (iii) | $a = (-)\omega^2 y$ **(1)**<br>$a = (-)1{\cdot}3^2 \times (-)4{\cdot}0$ **(1)**<br>$a = (-)6{\cdot}8\,\text{m}\,\text{s}^{-2}$ **(1)** | **3** | Accept:<br>7<br>6·8<br>6·76<br>6·760 |

| Question | | | Expected response | Max mark | Additional guidance |
|---|---|---|---|---|---|
| | | (iv) | Sine shape graph for one period of oscillation from $t = 0\text{s}$ to $t = 5\text{s}$ (1)<br>$v_{max} = \pm\omega\sqrt{(A^2 - y^2)}$ (1)<br>$v_{max} = \pm 1{\cdot}3 \times \sqrt{(4{\cdot}0^2 - 0^2)}$<br>$v_{max} = \pm 5{\cdot}2\,\text{m s}^{-1}$ (1) | 3 | Award a maximum of 2 marks if the labels, units or origin is/are missing. |
| | | (v) | $E_p = \frac{1}{2}m\omega^2 y^2$ (1)<br>$E_p = 0{\cdot}5 \times 85 \times 1{\cdot}3^2 \times 4{\cdot}0^2$ (1)<br>$E_p = 1{\cdot}1 \times 10^3\,\text{J}$ (1) | 3 | Accept:<br>$1 \times 10^3$<br>$1{\cdot}1 \times 10^3$<br>$1{\cdot}15 \times 10^3$<br>$1{\cdot}149 \times 10^3$ |

| Question | | | Expected response | Additional guidance |
|---|---|---|---|---|
| | (b) | | The whole candidate response should first be read to establish its overall quality in terms of accuracy and relevance to the problem/situation presented. There may be strengths and weaknesses in the candidate response: assessors should focus as far as possible on the strengths, taking account of weaknesses (errors or omissions) only where they detract from the overall answer in a significant way, which should then be taken into account when determining whether the response demonstrates reasonable, limited or no understanding.<br><br>Assessors should use their professional judgement to apply the guidance below to the wide range of possible candidate responses. | This open-ended question requires comment on possible reasons for discrepancies in assuming the SHM model. Candidate responses may include one or more of: snowboarder going too far; not a semicircle; movement down the half pipe; additional force caused by snowboarder or other relevant ideas/concepts. |
| | | | **3 marks:** The candidate has demonstrated a **good** conceptual understanding of the physics involved, providing a logically correct response to the problem/situation presented.<br><br>This type of response might include a statement of principle(s) involved, a relationship or equation, and the application of these to respond to the problem/situation.<br><br>This does not mean the answer has to be what might be termed an "excellent" answer or a "complete" one. | In response to this question, a **good** understanding might be demonstrated by a candidate response that:<br>• makes a judgement on suitability based on one relevant physics idea/concept, in a **detailed/developed** response that is **correct or largely correct** (any weaknesses are minor and do not detract from the overall response), **OR**<br>• makes judgement(s) on suitability based on a range of relevant physics ideas/concepts, in a response that is **correct or largely correct** (any weaknesses are minor and do not detract from the overall response), **OR**<br>• otherwise demonstrates a good understanding of the physics involved. |

| Question | | | Expected response | Additional guidance |
|---|---|---|---|---|
| | | | **2 marks:** The candidate has demonstrated a **reasonable** understanding of the physics involved, showing that the problem/situation is understood.<br><br>This type of response might make some statement(s) that is/are relevant to the problem/situation, for example, a statement of relevant principle(s) or identification of a relevant relationship or equation. | In response to this question, a **reasonable** understanding might be demonstrated by a candidate response that:'<br>• makes a judgement on suitability based on one or more relevant physics idea(s)/concept(s), in a response that is **largely correct** but has **weaknesses** which detract to a small extent from the overall response, **OR**<br>• otherwise demonstrates a reasonable understanding of the physics involved. |
| | | | **1 mark:** The candidate has demonstrated a **limited** understanding of the physics involved, showing that a little of the physics that is relevant to the problem/situation is understood.<br><br>The candidate has made some statement(s) that is/are relevant to the problem/situation. | In response to this question, a **limited** understanding might be demonstrated by a candidate response that:<br>• makes a judgement on suitability based on one or more relevant physics idea(s)/concept(s), in a response that has **weaknesses** which detract to a large extent from the overall response, **OR**<br>• otherwise demonstrates a limited understanding of the physics involved. |
| | | | **0 marks**: The candidate has demonstrated **no** understanding of the physics that is relevant to the problem/situation.<br><br>The candidate has made no statement(s) that is/are relevant to the problem/situation. | Where the candidate has *only* demonstrated knowledge and understanding of physics **that is not relevant to the problem/situation presented,** 0 marks should be awarded. |

| Question | | | Expected response | Max mark | Additional guidance |
|---|---|---|---|---|---|
| 6. | (a) | | Electrostatic force between the nucleus/proton and the electron. **(1)** | 1 | Any other forces shown 0 marks. |
| | (b) | (i) | (Electrostatic force = centripetal force)<br>$\frac{Q_1Q_2}{4\pi\varepsilon_0 r^2} = \frac{mv^2}{r}$ **(1)**<br>$\frac{e^2}{4\pi\varepsilon_0 r^2} = \frac{mv^2}{r}$<br>$mv^2 = \frac{e^2}{4\pi\varepsilon_0 r}$<br>$\frac{1}{2}mv^2 = \frac{e^2}{8\pi\varepsilon_0 r}$ **(1)**<br>$E_k = \frac{e^2}{8\pi\varepsilon_0 r}$ | 2 | Equations must be shown from Relationships Sheet to gain any marks.<br><br>If final line is not shown then maximum of 1 mark only can be awarded. |

| Question | | | Expected response | Max mark | Additional guidance |
|---|---|---|---|---|---|
| | | (ii) | $E_k = \frac{e^2}{8\pi\varepsilon_0 r}$<br>$E_k = \frac{(1{\cdot}60\times10^{-19})^2}{8\pi\times8{\cdot}85\times10^{-12}\times0{\cdot}21\times10^{-9}}$ **(1)**<br>$E_k = 5{\cdot}5\times10^{-19}$ J **(1)** | 2 | Accept:<br>$5\times10^{-19}$<br>$5{\cdot}5\times10^{-19}$<br>$5{\cdot}48\times10^{-19}$<br>$5{\cdot}483\times10^{-19}$<br>If $9\times10^9$ used, then accept 5, 5·5, 5·49, 5·486 |
| | (c) | | $mvr = \frac{nh}{2\pi}$ **(1)**<br>$4{\cdot}22\times10^{-34} = \frac{n\times6{\cdot}63\times10^{-34}}{2\pi}$ **(1)**<br>$n = 4$ **(1)(must be integer)** | 3 | |
| | (d) | (i) | $\Delta x\Delta p_x \geq \frac{h}{4\pi}$ **(1)**<br>$\Delta x\times1{\cdot}5\times10^{-26} \geq \frac{6{\cdot}63\times10^{-34}}{4\pi}$ **(1)**<br>minimum $\Delta x = 3{\cdot}5\times10^{-9}$ m **(1)** | 3 | Accept:<br>$4\times10^{-9}$,<br>$3{\cdot}52\times10^{-9}$,<br>$3{\cdot}517\times10^{-9}$ |
| | | (ii) | $\Delta t\Delta E \geq \frac{h}{4\pi}$ **(1)**<br>If $\Delta$t is small then $\Delta$E is large. **(1)**<br>Therefore the largest possible energy of the electron may be big enough to overcome the repulsion and cross the gap. **(1)**<br>**OR**<br>$\Delta x\Delta p \geq \frac{h}{4\pi}$ **(1)**<br>If the momentum is measured with a small uncertainty, the uncertainty in the position of the electron is large enough **(1)** for the electron to exist on the other side of the gap **(1)**. | 3 | |
| **7.** | (a) | | Reflected wave interferes with transmitted wave (to produce points of destructive and constructive interference). **(1)** | 1 | |
| | (b) | | Antinode (constructive), high energy, so melted spots. **(1)**<br>Node (destructive), low energy, so no melting. **(1)** | 2 | |
| | (c) | | $4\times\frac{1}{2}\lambda = 0{\cdot}264$ m **(1)**<br>$\lambda = 0{\cdot}132$ m **(1)** | 2 | |

| Question | | | Expected response | Max mark | Additional guidance |
|---|---|---|---|---|---|
| | (d) | | $v = f\lambda$ (1)<br>$= 2\cdot45 \times 10^9 \times 0\cdot132$ (1)<br>$= 3\cdot234 \times 10^8\,\text{m s}^{-1}$<br>$= 3\cdot23 \times 10^8\,\text{m s}^{-1}$ | 2 | If final answer is not shown then maximum of 1 mark only can be awarded. |
| | (e) | | The range of the results is small, so the results are precise. (1)<br>The difference between the mean value of the results and the accepted value of c is larger than the range, so the results are not accurate. (1) | 2 | |
| 8. | (a) | | A series of bright and dark spots. | 1 | Accept fringes. |
| | (b) | | $p = mv$<br>$p = 1\cdot20 \times 10^{-24} \times 220$<br>$= 2\cdot64 \times 10^{-22}\,(\text{kg m s}^{-1})$<br>$\lambda = \frac{h}{p}$ **(1)(for both formulae)**<br>$= \frac{6\cdot63 \times 10^{-34}}{2\cdot64 \times 10^{-22}}$ **(1)(for both substitutions)**<br>$= 2\cdot5 \times 10^{-12}\,\text{m}$ (1)<br>Estimate in the range of $10^{-12}$ to $10^{-9}$ (1) | 4 | Statement of value of slit separation must be distinct from value of $\lambda$. |
| 9. | (a) | | 3 | 1 | |
| | (b) | | $c = \frac{1}{\sqrt{\mu_0 \varepsilon_0}}$ (1)<br>$= \frac{1}{\sqrt{8\cdot93 \times 10^{-12} \times 1\cdot32 \times 10^{-6}}}$ (1)<br>$= 2\cdot91 \times 10^8\,\text{m s}^{-1}$ (1) | 3 | The answer must be consistent with (a) in terms of significant figures.<br>If not consistent then maximum of 2 marks only can be awarded. |
| | (c) | (i) | %uncert in $\mu_0 = \pm\frac{5 \times 10^{-8}}{1\cdot32 \times 10^{-6}} \times 100$<br>$= \pm3\cdot8\%$ (1)<br>%uncert in $\varepsilon_0 = \pm\frac{7 \times 10^{-14}}{8\cdot93 \times 10^{-12}} \times 100$<br>$= \pm0\cdot8\%$ (1)<br>Uncertainty in $\mu_0$ more significant (1) | 3 | |
| | | (ii) | uncert in $\frac{1}{\sqrt{\mu_0 \varepsilon_0}} = \pm\frac{1}{2} \times \frac{3\cdot8}{100} \times 2\cdot91 \times 10^8$ (1)<br>$= \pm6 \times 10^6\,\text{m s}^{-1}$ (1) | 2 | If a candidate combines both uncertainties correctly full marks may be awarded. |

| Question | | | Expected response | Max mark | Additional guidance |
|---|---|---|---|---|---|
| 10. | (a) | (i) | Force acting per unit positive charge. | 1 | |
| | | (ii) | $\frac{Q_1}{4\pi\varepsilon_0 r_1^2} = \frac{Q_2}{4\pi\varepsilon_0 r_2^2}$ (1)<br>$\frac{-8\cdot0\times10^{-9}}{4\pi\varepsilon_0(0\cdot4)^2} = \frac{Q_S}{4\pi\varepsilon_0(0\cdot2)^2}$ (1)<br>$Q_2 = \frac{-8\cdot0\times10^{-9}\times(0\cdot2)^2}{(0\cdot4)^2}$<br>$Q_2 = -2\cdot0\times10^{-9}$ C | 2 | If final answer is not shown then maximum of 1 mark can be awarded. |
| | | (iii) | $V_1 = \frac{Q}{4\pi\varepsilon_0 r}$ (1)<br>$V_1 = \frac{-8\cdot0\times10^{-9}}{4\times\pi\times8\cdot85\times10^{-12}\times0\cdot40}$ (1)<br>$V_1 = -180$ V (1)<br>$V_2 = \frac{-2\cdot0\times10^{-9}}{4\times\pi\times8\cdot85\times10^{-12}\times0\cdot20}$<br>$V_2 = -90$ V (1)<br>Potential at X = –180 – 90 = –270 V (1) | 5 | $V_1 = -179\cdot84$<br>$V_2 = -89\cdot92$ |
| | (b) | (i) | $V_1 = \frac{-8\cdot0\times10^{-9}}{4\times\pi\times8\cdot85\times10^{-12}\times0\cdot50}$<br>$V_1 = -140$ V (1)<br>$V_2 = \frac{-2\cdot0\times10^{-9}}{4\times\pi\times8\cdot85\times10^{-12}\times0\cdot50}$<br>$V_2 = -36$ V (1)<br>Potential at P = –140 – 36 = –176 V = –180 V (1) | 3 | Accept:<br>$V_1 = -144$ V<br>Accept:<br>–176 V |
| | | (ii) | Potential difference = –180 – (–270) = 90 (V) (1)<br>$E = QV$ (1)<br>$= 1\cdot0\times10^{-9}\times90$ (1)<br>$= 9\cdot0\times10^{-8}$ J (1) | 4 | Accept potential difference = –176 – (–270) = 94 V leading to $E = 9\cdot4\times10^{-8}$ J |

| Question | | | Expected response | Additional guidance |
|---|---|---|---|---|
| 11. | | | The whole candidate response should first be read to establish its overall quality in terms of accuracy and relevance to the problem/ situation presented. There may be strengths and weaknesses in the candidate response: assessors should focus as far as possible on the strengths, taking account of weaknesses (errors or omissions) only where they detract from the overall answer in a significant way, which should then be taken into account when determining whether the response demonstrates reasonable, limited or no understanding.<br><br>Assessors should use their professional judgement to apply the guidance below to the wide range of possible candidate responses. | This open-ended question requires comment on the statement "things on a small scale behave nothing like things on a large scale". Candidate responses may include one or more of: macroscopic/microscopic, duality; uncertainty; double slit; failure of Newtonian rules in the atomic world; intuition applies to large objects or other relevant ideas/ concepts. |
| | | | **3 marks:** The candidate has demonstrated a **good** conceptual understanding of the physics involved, providing a logically correct response to the problem/ situation presented.<br><br>This type of response might include a statement of principle(s) involved, a relationship or equation, and the application of these to respond to the problem/situation.<br><br>This does not mean the answer has to be what might be termed an "excellent answer" or a "complete" one. | In response to this question, a **good** understanding might be demonstrated by a candidate response that:<br>• makes a judgement on suitability based on one relevant physics idea/concept, in a **detailed/developed** response that is **correct or largely correct** (any weaknesses are minor and do not detract from the overall response), **OR**<br>• makes judgement(s) on suitability based on a range of relevant physics ideas/ concepts, in a response that is **correct or largely correct** (any weaknesses are minor and do not detract from the overall response), **OR**<br>• otherwise demonstrates a good understanding of the physics involved. |
| | | | **2 marks:** The candidate has demonstrated a **reasonable** understanding of the physics involved, showing that the problem/ situation is understood.<br><br>This type of response might make some statement(s) that is/ are relevant to the problem/ situation, for example, a statement of relevant principle(s) or identification of a relevant relationship or equation. | In response to this question, a **reasonable** understanding might be demonstrated by a candidate response that:<br>• makes a judgement on suitability based on one or more relevant physics idea(s)/ concept(s), in a response that is **largely correct** but has **weaknesses** which detract to a small extent from the overall response, **OR**<br>• otherwise demonstrates a reasonable understanding of the physics involved. |

| Question | | | Expected response | Additional guidance |
|---|---|---|---|---|
| | | | **1 mark:** The candidate has demonstrated a **limited** understanding of the physics involved, showing that a little of the physics that is relevant to the problem/situation is understood.<br><br>The candidate has made some statement(s) that is/are relevant to the problem/situation. | In response to this question, a **limited** understanding might be demonstrated by a candidate response that:<br>• makes a judgement on suitability based on one or more relevant physics idea(s)/concept(s), in a response that has **weaknesses** which detract to a large extent from the overall response, **OR**<br>• otherwise demonstrates a limited understanding of the physics involved. |
| | | | **0 marks:** The candidate has demonstrated **no** understanding of the physics that is relevant to the problem/situation.<br><br>The candidate has made no statement(s) that is/are relevant to the problem/situation. | Where the candidate has *only* demonstrated knowledge and understanding of physics **that is not relevant to the problem/situation presented,** 0 marks should be awarded. |

| Question | | | Expected response | | Max mark | Additional guidance |
|---|---|---|---|---|---|---|
| **12.** | (a) | (i) | $X_C = \frac{1}{2\pi fC}$<br>$= \frac{1}{2 \times \pi \times 65 \times 5{\cdot}0 \times 10^{-6}}$<br>$= 490\,\Omega$ | (1)<br><br>(1) | 3 | Accept:<br>500<br>490<br>489·7 |
| | | (ii) | $I_{rms} = \frac{V_{rms}}{X_C}$<br>$= \frac{15}{490}$<br>$= 3{\cdot}1 \times 10^{-2}$ A | (1)<br>(1)<br>(1) | 3 | Accept:<br>3<br>3·1<br>3·06<br>3·061 |
| | (b) | (i) | Plot $X_C$ against 1/f<br>Labels (quantities and units) and scale<br>Points plotted correctly<br>Correct best fit line | (1)<br>(1)<br>(1)<br>(1) | 4 | Non-linear scale a maximum of 1 mark is available.<br>Allow ± half box tolerance when plotting points. |
| | | (ii) | Gradient of best fit line<br>Gradient $= \frac{1}{2\pi C}$<br>**OR**<br>$C = \frac{1}{(2\pi \times \text{gradient})}$<br><br>Final value of C | (1)<br><br><br>(1)<br><br>(1) | 3 | If candidates use data points not on their line of best fit, then maximum of 1 mark is available.<br>A representative gradient value of $3{\cdot}13 \times 10^{7}$ gives a capacitance of $5{\cdot}08 \times 10^{-9}$ F.<br>Final value of C must be consistent with candidate's value for gradient. |

## ADVANCED HIGHER PHYSICS MODEL PAPER 1

| Question | | | Expected response | Max mark |
|---|---|---|---|---|
| 1. | (a) | (i) | $s = ut + \frac{1}{2}at^2$<br>$(u = 0)$<br>$s = \frac{1}{2}at^2$<br>$a = \frac{2s}{t^2}$ | 1 |
| | | (ii) | Mean $t = 2{\cdot}45$ (s) **(1)**<br>$a = \frac{2 \times 3{\cdot}54}{2{\cdot}45^2}$<br>$a = 1{\cdot}18\,\text{m s}^{-2}$ **(1)** | 2 |
| | | (iii) | Uncertainty $= \pm\left(\frac{t_{max} - t_{min}}{n}\right)$<br>Uncertainty $= \pm\left(\frac{2{\cdot}65 - 2{\cdot}29}{6}\right)$ **(1)**<br>Uncertainty $= \pm\, 0{\cdot}06$ s $(= \pm 2{\cdot}45\%)$ **(1)** | 2 |
| | | (iv) | **Time**<br>% Uncertainty $= \pm\left(\frac{0{\cdot}06}{2{\cdot}45} \times 100\right) = \pm 2{\cdot}45\%$<br>% Uncertainty in $t^2 = \pm 4{\cdot}9\%$ **(1)**<br>**Distance**<br>% Uncertainty $= \pm\left(\frac{0{\cdot}01}{3{\cdot}54} \times 100\right) = \pm 0{\cdot}28\%$ **(1)**<br>∴ ignore<br>Uncertainty in $a = \pm 4{\cdot}9\%$ **(1)** | 3 |
| | | (v) | Uncertainty $= \pm\left(\frac{1{\cdot}18 \times 4{\cdot}9}{100}\right) = \pm 0{\cdot}06\ (\text{m s}^{-2})$ **(1)**<br>$a = (1{\cdot}18 \pm 0{\cdot}06)\ (\text{m s}^{-2})$ **(1)** | 2 |
| | (b) | (i) | $a = \frac{v^2}{r}$ **(1)**<br>$a = \frac{36}{4}$ **(1)**<br>$a = 9{\cdot}0\,\text{m s}^{-2}$ **(1)** | 3 |
| | | (ii) | $F_{radial} = \frac{mv^2}{r}$ **(1)**<br>$F_{radial} = 2{\cdot}5 \times 9$<br>$F_{radial} = 22{\cdot}5$ (N) **(1)**<br>Frictional force is sufficient to provide this. **(1)** | 3 |
| | (c) | | Component of the reaction force now acts radially, **(1)**<br>the central force is increased. **(1)** | 2 |
| 2. | (a) | | $I = mr^2$ | 1 |
| | (b) | | $I = 1{\cdot}5 \times 0{\cdot}20^2$<br>$I = 0{\cdot}060\,\text{kg m}^2$ (given) | 1 |
| | (c) | (i) | $T = Fr = (25 \times 4 \times 10^{-3})$ **(1)**<br>Resultant $T = (25 \times 4 \times 10^{-3}) - 0{\cdot}070$ **(1)**<br>Resultant $T = 0{\cdot}030\,\text{N m}$ **(1)** | 3 |
| | | (ii) | $T = I\alpha$ **(1)**<br>$0{\cdot}030 = 0{\cdot}060 \times \alpha$ **(1)**<br>$\alpha = 0{\cdot}50\,\text{rad s}^{-2}$ **(1)** | 3 |
| | | (iii) | No. of revolutions $= \frac{\text{cord length}}{\text{circumference}}$ **(1)**<br>No. of revolutions $= \frac{0{\cdot}5}{(2 \times \pi \times 4 \times 10^{-3})}$ **(1)**<br>No. of revolutions = 19·9<br>$\theta = \text{N}^\circ$ of revolutions $\times 2\pi$ **(1)**<br>$\theta = 19{\cdot}9 \times 2\pi$<br>$\theta = 125$ rad | 3 |
| | | (iv) | $\omega^2 = \omega_0^{\,2} + 2\alpha\theta$ **(1)**<br>$\omega^2 = 0 + 2 \times 0{\cdot}5 \times 125$ **(1)**<br>$\omega = 11\,\text{rad s}^{-1}$ **(1)** | 3 |
| | | (v) | $\alpha = \frac{T}{I}$ **(1)**<br>$\alpha = (-)\frac{0{\cdot}07}{0{\cdot}06}$<br>$\alpha = (-)1{\cdot}2\ (\text{rad s}^{-2})$ **(1)**<br>$\alpha = \frac{\omega - \omega_0}{t}$ **(1)**<br>$t = \frac{4{\cdot}2 - 11}{-1{\cdot}2}$<br>$t = 5{\cdot}7$ s **(1)** | 4 |

| Question | | | Expected response | Max mark |
|---|---|---|---|---|
| 3. | (a) | | $F=\frac{GMm}{r^2}$ (1)<br>$F=$<br>$\frac{6{\cdot}67\times10^{-11}\times6{\cdot}0\times10^{24}\times7{\cdot}3\times10^{22}}{(3{\cdot}84\times10^{8})^2}$ (1)<br>$F=2{\cdot}0\times10^{20}$ N (given) | 2 |
| | (b) | | $F=\frac{mv^2}{r}$ (1)<br>$2{\cdot}0\times10^{20}=\frac{7{\cdot}3\times10^{22}\times v^2}{3{\cdot}84\times10^{8}}$ (1)<br>$v=1{\cdot}0\times10^{3}$ m s$^{-1}$ (1) | 3 |
| | (c) | | $E_P=\frac{-GMm}{r}$ (1)<br>$E_P=$<br>$\frac{-6{\cdot}67\times10^{-11}\times6{\cdot}0\times10^{24}\times7{\cdot}3\times10^{22}}{3{\cdot}84\times10^{8}}$ (1)<br>$E_P=-7{\cdot}6\times10^{28}$ J (1) | 3 |
| | (d) | | $E_T=E_K+E_P$ (1)<br>**OR**<br>$E_T=\frac{1}{2}mv^2-7{\cdot}6\times10^{28}$ J (1)<br>$E_T=\frac{1}{2}\times7{\cdot}3\times10^{22}\times(1{\cdot}0\times10^{3})^2$<br>$-7{\cdot}6\times10^{28}$ J (1)<br>$E_T=-4{\cdot}0\times10^{28}$ J (1)<br>**OR**<br>$E_T=\frac{-GMm}{2r}$ (1)<br>Sub (1)<br>$E_T=-3{\cdot}8\times10^{28}$ J (1) | 3 |
| 4. | (a) | | Massive objects curve spacetime (1)<br>Other objects follow a curved path through this (distorted) spacetime (1) | 2 |
| | (b) | (i) | Curved path around massive object<br>(curve must be shown)<br>Apparent position (ii)<br>Earth<br>Massive object<br>(i) Star (i)<br>Apparent position (ii) | 1 |
| | | (ii) | Light beam from apparent position to observer | 1 |
| | (c) | | B (1)<br>Time passes more slowly at lower altitudes (in a gravitational field).<br>**OR**<br>Lower gravitational field strength at higher altitude. (1) | 2 |
| 5. | | | **Open-ended question**<br>Answer might include:<br>• Attendant **appears** to be thrown outwards in the rotating frame of reference.<br>• Tom's view is from a rotating position<br>• Looking at the situation from above the attendant will be moving at 90° to the radius i.e. travels along the tangent to the circle.<br>• This is due to a lack of centripetal force acting on the attendant to maintain his circular motion.<br>• There is no outward (centrifugal) force acting on the attendant. | 3 |

| Question | | | Expected response | Max mark |
|---|---|---|---|---|
| 6. | (a) | (i) | Electron diffraction<br>**OR**<br>Electron microscopy | 1 |
| | | (ii) | Deflection tube<br>**OR**<br>Photoelectric effect<br>**OR**<br>Similar answers | 1 |
| | (b) | (i) | $\lambda = \frac{h}{mv}$ **(1)**<br>$\lambda = \frac{6{\cdot}63 \times 10^{-34}}{0{\cdot}06 \times 55}$ **(1)**<br>$\lambda = 2{\cdot}0 \times 10^{-34}$ m **(1)** | 3 |
| | | (ii) | $\lambda$ too small be observed | 1 |
| 7. | (a) | (i) | $\lambda = \frac{h}{p}$<br>$\lambda = \frac{h}{mv}$<br>$\lambda = \frac{6{\cdot}63 \times 10^{-34}}{9{\cdot}11 \times 10^{-31} \times 3{\cdot}2 \times 10^{6}}$ **(1)**<br>$\lambda = 2{\cdot}3 \times 10^{-10}$ (m) **(1)**<br>$\Delta x \Delta p \geq \frac{h}{4\pi}$ **(1)**<br>$2{\cdot}3 \times 10^{-10} \Delta p \geq \frac{6{\cdot}63 \times 10^{-34}}{4\pi}$<br>$\Delta p \geq 2{\cdot}3 \times 10^{-25}$ kg m s$^{-1}$ (Ns) **(1)** | 4 |
| | | (ii) | $\lambda$ reduced (or $f$ increased) for X-rays,<br>**OR**<br>greater $E$ transferred **(1)**<br>$\Delta x$ reduced for X-rays **(1)**<br>Since $\Delta x \Delta p \geq \frac{h}{4\pi}$<br>$\Delta p$ increases **(1)** | 3 |
| | (b) | | Since $\Delta E \Delta t \geq \frac{h}{4\pi}$ **(1)**<br>Borrowing energy for a short period of time allows particles to escape **(1)** | 2 |

| Question | | | Expected response | Max mark |
|---|---|---|---|---|
| 8. | (a) | | $T^2 = \frac{4\pi^2 l}{g}$ **(1)**<br>Gradient of graph $= \frac{4\pi^2}{g}$ **(1)**<br>Gradient of graph $= \frac{y_2 - y_1}{x_2 - x_1}$ **(1)**<br>Gradient of graph = 3·7<br>$g = \frac{4\pi^2}{3{\cdot}7}$<br>$g = 10{\cdot}7$ N kg$^{-1}$ **(1)** | 4 |
| | (b) | | % uncertainty in gradient $= \frac{0{\cdot}12}{3{\cdot}7} \times 100$ **(1)**<br>% uncertainty in gradient = (±) 3·2% **(1)**<br>Absolute uncertainty in g $= \frac{3{\cdot}2}{100} \times 10{\cdot}7$<br>Absolute uncertainty in g = (±) 0·3 N kg$^{-1}$ **(1)** | 3 |
| | (c) | | Error bars will indicate the precision of each point. **(1)**<br>This will enable the student to highlight possible improvements. **(1)** | 2 |
| | (d) | (i) | Estimate the random, calibration and scale reading uncertainty for each mean value of $T$.<br>Combine these using the Pythagorean relationship to give uncertainty in the mean of $T$. **(1)**<br>Calculate the mean value of $T$ for each set of readings.<br>Calculate the percentage uncertainty in $T$. **(1)**<br>Double this to find the percentage uncertainty in $T^2$. **(1)**<br>Convert this to the absolute uncertainty in $T^2$. **(1)** | 4 |

| Question | | | Expected response | Max mark |
|---|---|---|---|---|
| | | (ii) | Estimate the random, calibration and scale reading uncertainty for each value of $l$. (1)<br>Combine these using the Pythagorean relationship to give uncertainty in length. (1) | 2 |
| 9. | (a) | (i) | By comparison with "standard" (1)<br>$y = A\sin 2\pi\left(ft - \frac{x}{\lambda}\right)$<br>$2\pi f = 12$ (1)<br>$f = 1{\cdot}9\,\text{Hz}$ (1) | 3 |
| | | (ii) | $\frac{2\pi}{\lambda} = 0{\cdot}50$ (1)<br>$\lambda = 13\,\text{m}$ (1) | 2 |
| | (b) | (i) | $\Delta x = 1\,\text{m}$ (1)<br>$\phi = \frac{\Delta x}{\lambda} \times 2\pi$ (1)<br>$\phi = \frac{1}{13} \times 2\pi$ (1)<br>$\phi = 0{\cdot}48\,\text{rad}$ (1) | 4 |
| | | (ii) | $v = f\lambda$ (1)<br>$v = 1{\cdot}9 \times 13$<br>$v = 24{\cdot}7\,\text{m s}^{-1}$ (1)<br>$t = \frac{d}{v}$<br>$t = \frac{1}{24{\cdot}7}$<br>$t = 0{\cdot}04\,\text{s}$ (1) | 3 |
| | (c) | | $y = A\sin(12t + 0{\cdot}5x)$ (1)<br>where $A < 8$ (1) | 2 |
| 10. | (a) | | Division of amplitude. | 1 |
| | (b) | (i) | Ray 1 – phase change of $\pi$. | 1 |
| | | (ii) | Ray 2 – no phase change. | 1 |
| | (c) | | For constructive interference<br>Optical Path Difference $= \left(m + \frac{1}{2}\right)\lambda$ (1)<br>2nd $= \frac{\lambda}{2}$ ($m = 0$ for min thickness) (1)<br>$d = \frac{\lambda}{4n}$ | 2 |
| | (d) | | $d = \frac{\lambda}{4n}$<br>$d = \frac{550 \times 10^{-9}}{4 \times 2{\cdot}4}$ (1)<br>$d = 5{\cdot}73 \times 10^{-8}\,\text{m}$ (1) | 2 |
| | (e) | | Use another thin layer of glass, place on top (1)<br>and then coat this surface with the same thickness. (1) | 2 |
| 11. | | | **Open-ended question**<br>Answer might include:<br>• The colours of light are being split up by the screen.<br>• The TV screen must be acting as a diffraction grating for reflected light.<br>• Screen contains closely spaced vertical/ horizontal lines.<br>• Colours produced by constructive interference of reflected light.<br>• Spacing of colours depends on angle of view, separation of lines on screen.<br>• Effect caused by division of wavefront.<br>• Although incorrect, there could be a thin film explanation given, where reflected light interferes. This would be division of amplitude. | 3 |
| 12. | (a) | | $F = qvB$ (1)<br>$5 \times 10^{-11} = 3{\cdot}2 \times 10^{-19} \times v \times 6{\cdot}8$ (2)<br>(1 sub, 1 for charge on $\alpha$ particle).<br>$v = 2{\cdot}3 \times 10^{7}\,\text{m s}^{-1}$ (given) | 3 |

| Question | | | Expected response | Max mark |
|---|---|---|---|---|
| | (b) | | $v = \frac{E}{B}$ **(1)**<br>$2{\cdot}3 \times 10^7 = \frac{E}{6{\cdot}8}$ **(1)**<br>$E = 1{\cdot}6 \times 10^8\,\text{V m}^{-1}$ **(1)**<br>**OR**<br>$E = \frac{F}{Q}$ **(1)**<br>$E = \frac{5{\cdot}0 \times 10^{-11}}{3{\cdot}2 \times 10^{-19}}$ **(1)**<br>$E = 1{\cdot}6 \times 10^8\,\text{N C}^{-1}$ **(1)** | 3 |
| | (c) | | $F = \frac{mv^2}{r}$ **(1)**<br>$5{\cdot}0 \times 10^{-11} = \frac{6{\cdot}645 \times 10^{-27} \times (2{\cdot}3 \times 10^7)^2}{r}$ **(1)**<br>$r = 0{\cdot}070$ (m) **(1)**<br>alpha particle hits at position B or 0·14 m **(1)** | 4 |
| | (d) | | Electron will be deflected in the opposite direction<br>due to opposite charge **(1)**<br>radius of semicircle smaller **(1)**<br>due to (much) **smaller** mass or greater $\frac{q}{m}$ **(1)** | 3 |
| **13.** | (a) | (i) | Circuit must be able to make required measurements as shown or zero marks.<br>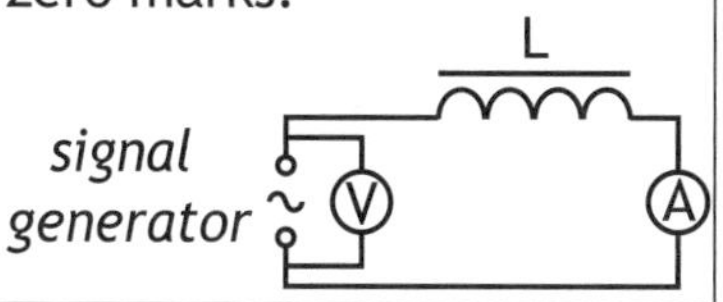 | 2 |
| | | (ii) | $k$ values are<br>5·9 6·1 6·1 5·8 6·0<br>All $k$ values correct **(1)**<br>$I$ inversely proportional to $f$ **(1)** | 2 |
| | (b) | (i) | $V_s = 20$ (V) **(1)**<br>$V_L = 9\,\text{V}$ **(1)**<br>$V_R = 20 - 9$<br>$V_R = 11\,\text{V}$ **(1)** | 3 |

| Question | | | Expected response | Max mark |
|---|---|---|---|---|
| | | (ii) | $E = -L\frac{dI}{dt}$ **(1)**<br>$-4{\cdot}2 = -3 \times \frac{dI}{dt}$ **(1)**<br>$\frac{dI}{dt} = 1{\cdot}4\,\text{A s}^{-1}$ **(1)** | 3 |
| | | (iii) | Rate of change of current or rate of change of magnetic field is at its maximum. | 1 |
| | (c) | (i) | For equations and for equality<br>$2\pi f_0 L = \frac{1}{2\pi f_0 C}$ | 1 |
| | | (ii) | $f_0 = \frac{1}{2\pi\sqrt{LC}}$<br>$f_0 = \frac{1}{2\pi\sqrt{2{\cdot}2 \times 10^{-3} \times 4{\cdot}7 \times 10^{-6}}}$ **(1)**<br>$f_0 = 1600\,\text{Hz}$ **(1)** | 2 |
| | | (iii) | $4.0\,\Omega$ | 1 |

## ADVANCED HIGHER PHYSICS MODEL PAPER 2

| Question | | | Expected response | Max mark |
|---|---|---|---|---|
| 1. | (a) | | $\omega = \frac{\theta}{t}$<br>$\omega = \frac{3 \cdot 1}{4 \cdot 5}$<br>$\omega = 0 \cdot 69\,(\text{m s}^{-1})$ (1)<br>$v = r\omega$ (1)<br>$v = 0 \cdot 148 \times 0 \cdot 69$<br>$v = 0 \cdot 10\,\text{m s}^{-1}$ (1) | 3 |
| | (b) | | $\%\Delta\theta = \frac{0 \cdot 1}{3 \cdot 1} \times 100 = 3 \cdot 2\%$ (1)<br>$\%\Delta t = \frac{0 \cdot 1}{4 \cdot 5} \times 100 = 2 \cdot 2\%$ (1)<br>$\%\Delta r = \frac{0 \cdot 001}{0 \cdot 148} \times 100$<br>$= 0 \cdot 68\%$ (1)<br>$\%\Delta\omega = \sqrt{\%\Delta\theta^2 + \%\Delta t^2 + \%\Delta r^2}$<br>$\%\Delta\omega = 3 \cdot 9\,(\%)$ (1) | 4 |
| 2. | (a) | (i) | $I_{rod} = \frac{1}{3} ml^2$ (1)<br>$I_{rod} = \frac{1}{3} \times 0 \cdot 040 \times 0 \cdot 30^2$ (1)<br>$I_{rod} = 1 \cdot 2 \times 10^{-3}\,\text{kg m}^2$ (1) | 3 |
| | | (ii) | $I_{wheel} = (5 \times I_{rod}) + m_{(rim)} r^2$ (2)<br>$= (5 \times 1 \cdot 2 \times 10^{-3}) + (0 \cdot 24 \times 0 \cdot 30^2)$ (2)<br>$= 6 \times 10^{-3} + 0 \cdot 0216$<br>$= 0 \cdot 0276$<br>$= 0 \cdot 028\,(\text{kg m}^2)$ (given) | 4 |
| | (b) | (i) | $v = \omega r$ (1)<br>$19 \cdot 2 = \omega \times 0 \cdot 3$ (1)<br>$\omega = 64\,\text{rad s}^{-1}$ (1) | 3 |
| | | (ii)(A) | $\omega = \omega_0 + \alpha t$ (1)<br>$0 = 64 + \alpha \times 6 \cdot 7$ (1)<br>$\alpha = -9 \cdot 6\,\text{rad s}^{-2}$ (1) | 3 |
| | | (ii)(B) | $T = I \times \alpha$ (1)<br>$T = 0 \cdot 028 \times (-)9 \cdot 6$ (1)<br>$T = (-)0 \cdot 27\,\text{N m}$ (1) | 3 |
| 3. | (a) | | $(m)g = \frac{GM(m)}{R^2}$ (2)<br>$R^2 = \frac{6 \cdot 67 \times 10^{-11} \times 6 \cdot 4 \times 10^{23}}{3 \cdot 7}$<br>$R^2 = 1 \cdot 15 \times 10^{13}$<br>$R = 3 \cdot 4 \times 10^6\,\text{m}$ (1) | 3 |

| Question | | | Expected response | Max mark |
|---|---|---|---|---|
| | (b) | (i) | $\frac{GMm}{R^2} = m\omega^2 R$ (2)<br>$\frac{GM}{R^2} = \omega^2 R$ (1)<br>$\omega^2 = \frac{GM}{R^3}$<br>$\omega = \sqrt{\frac{GM}{R^3}}$ | 3 |
| | | (ii) | $\omega = \sqrt{\frac{GM}{R^3}}$<br>$\omega = \sqrt{\frac{6 \cdot 67 \times 10^{-11} \times 6 \cdot 4 \times 10^{23}}{(3 \cdot 4 \times 10^6 + 1 \cdot 7 \times 10^7)^3}}$<br>**(1) for top line; (1) for bottom line**<br>$\omega = 7 \cdot 1 \times 10^{-5}\,\text{rad s}^{-1}$ (1) | 3 |
| | | (iii) | $T = \frac{2\pi}{\omega}$ (1)<br>$T = \frac{2 \times 3 \cdot 14}{7 \cdot 1 \times 10^{-5}}$ (1)<br>$T = 8 \cdot 9 \times 10^4\,\text{s}$ (1) | 3 |
| | (c) | | Venus: $\frac{T^2}{R^3} = 3 \cdot 05 \times 10^{-7}$<br>Mars: $\frac{T^2}{R^3} = 3 \cdot 02 \times 10^{-7}$<br>Jupiter: $\frac{T^2}{R^3} = 3 \cdot 03 \times 10^{-7}$ (2)<br>Statement $\frac{T^2}{R^3} = \text{constant}$ (1)<br>**OR**<br>Draw graph of $T^2$ vs $R^3$ to give straight line through origin. (3) | 3 |
| 4. | (a) | (i) | $L = 1 \times 10^{-2}$ solar luminosities (from diagram) (1)<br>$L = 1 \times 10^{-2} \times 3 \cdot 9 \times 10^{26}$<br>$= 3 \cdot 9 \times 10^{24}\,(\text{W})$ (1) | 2 |
| | | (ii) | $T = 3000\,\text{K}$ (from diagram) (1)<br>$L = 4\pi r^2 \sigma T^4$ (1)<br>$3 \cdot 9 \times 10^{24} = 4\pi r^2 \times 5 \cdot 67 \times 10^{-8} \times 3000^4$<br>$r = 2 \cdot 6 \times 10^8\,\text{m}$<br>$r = 3 \times 10^8\,\text{m}$ 1 s.f. (1) | 3 |

| Question | | | Expected response | Max mark |
|---|---|---|---|---|
| | | (iii) | Difficult scale to read/ information from diagram can only be read to 1 s.f. | 1 |
| | (b) | (i) | $f_{peak} = \frac{2 \cdot 8 k_b T}{h}$<br>$T = 3000\,K$ (1)<br>$f_{peak} = \frac{2 \cdot 8 \times 1 \cdot 38 \times 10^{-23} \times 3000}{6 \cdot 63 \times 10^{-34}}$ (1)<br>$f_{peak} = 2 \times 10^{14}\,Hz$ (1) | 3 |
| | | (ii) | $v = f\lambda$ (1)<br>$3 \cdot 0 \times 10^8 = f \times 1 \cdot 9 \times 10^{-3}$<br>$f = 1 \cdot 6 \times 10^{11}$ (Hz) (1)<br>$f_{peak} = \frac{2 \cdot 8 k_b T}{h}$<br>$1 \cdot 6 \times 10^{11} = \frac{2 \cdot 8 \times 1 \cdot 38 \times 10^{-23} T}{6 \cdot 63 \times 10^{-34}}$ (1)<br>$T = 2 \cdot 7\,K$ (1) | 4 |
| | (c) | | $M_{black\ hole} = 2 \cdot 0 \times 10^{30} \times 1 \cdot 0 \times 10^{-10}$<br>$= 2 \cdot 0 \times 10^{20}$ (kg) (1)<br>$r_{Schwarzschild} = \frac{2GM}{c^2}$ (1)<br>$r_{Schwarzschild} = \frac{2 \times 6 \cdot 67 \times 10^{-11} \times 2 \cdot 0 \times 10^{20}}{(3 \cdot 0 \times 10^8)^2}$<br>$r_{Schwarzschild} = 3 \cdot 0 \times 10^{-7}\,m$ (1) | 3 |
| 5. | | | **Open-ended question**<br>Answer might include:<br>• To ensure the car has enough energy then $E_p = E_k$ (linear at top of loop) + heat, sound due to friction. $E_p = mgh$ where $h$ is the **difference** in height from the start to the **top** of the loop or $h$ = height from bottom minus $2r$ where $r$ = radius of the loop.<br>• Car should have enough speed at the bottom of the loop to complete the motion. The minimum speed required will ensure the weight of the car provides the central force at the top of the loop.<br>• At top of loop $mg = \frac{mv^2}{r}$, so required $v = \sqrt{gr}$<br>• As the car moves from the bottom of the loop $E_k$ is converted to $E_p$.<br>• $\frac{1}{2}mv^2 = mgh +$ **energy loss**<br>• Starting height, (mass) and speed of the cars would have to be accurately calculated.<br>• Frictional losses estimated.<br>• Allowance made so that the height easily overcomes energy losses.<br>• Passengers properly strapped into car, especially if upside down.<br>• Padding where required to reduce the effect of sudden changes in velocity (momentum) (Increasing the time in contact reduces the acting force).<br>• Cars attached to the rails so they do cannot fall off at the top of the loop. | 3 |
| 6. | (a) | (i) | $mvr = \frac{nh}{2\pi}$ (1)<br>$mvr = \frac{1 \times 6 \cdot 63 \times 10^{-34}}{2\pi}$ (1)<br>$= 1 \cdot 06 \times 10^{-34}\,kg\,m^2\,s^{-1}$ or $kg\,m^2\,rad\,s^{-1}$ or $J\,s$ (1) | 3 |

| Question | | | Expected response | Max mark |
|---|---|---|---|---|
| | | (ii) | $mv = \frac{nh}{2\pi r}$ (1)<br>$mv = \frac{1 \times 6{\cdot}63 \times 10^{-34}}{5{\cdot}3 \times 10^{-11} \times 2\pi}$ (1)<br>$mv = 2{\cdot}0 \times 10^{-24}\,\text{kg m s}^{-1}$ (1)<br>**OR**<br>$mv = \frac{L}{r}$ (1)<br>$mv = \frac{1{\cdot}06 \times 10^{-34}}{5{\cdot}3 \times 10^{-11}}$ (1)<br>$mv = 2{\cdot}0 \times 10^{-24}\,\text{kg m s}^{-1}$ (1) | 3 |
| | | (iii) | $\lambda = \frac{h}{p}$ (1)<br>$\lambda = \frac{6{\cdot}63 \times 10^{-34}}{2{\cdot}0 \times 10^{-24}}$ (1)<br>$\lambda = 3{\cdot}3 \times 10^{-10}\,\text{m}$ (1) | 3 |
| | (b) | (i) | The electrons would (spiral) inwards towards the nucleus.<br>**OR**<br>Orbit decays/decreases | 1 |
| | | (ii) | Quantum mechanics | 1 |
| 7. | (a) | (i) | Acceleration $\alpha$-(displacement)<br>**OR**<br>force $\alpha$ (displacement) and directed towards a fixed point | 1 |
| | | (ii) | $T = \frac{2\pi}{\omega}$ (1)<br>$T = \frac{2 \times 3{\cdot}14}{4{\cdot}3}$ (1)<br>$T = 1{\cdot}5\,\text{s}$ (or $T = 1{\cdot}46\,\text{s}$) (1) | 3 |
| | (b) | | $v_{max} = \omega A$ (1)<br>$v_{max} = 4{\cdot}3 \times 2 \times 10^{-2}$ (1)<br>$v_{max} = 8{\cdot}6 \times 10^{-2}\,\text{m s}^{-1}$ (1) | 3 |
| | (c) | | $E_{total} = E_k \text{ max}$<br>$E_{total} = \frac{1}{2}mv^2$ (1)<br>$E_{total} = 0{\cdot}5 \times 5{\cdot}0 \times 10^{-2} \times (8{\cdot}6 \times 10^{-2})^2$ (1)<br>$E_{total} = 1{\cdot}8 \times 10^{-4}\,\text{J}$ (1) | 3 |

| Question | | | Expected response | Max mark |
|---|---|---|---|---|
| | (d) | | $T = 2\pi\sqrt{\frac{L}{g}}$<br>$1{\cdot}5 = 2 \times 3{\cdot}14 \times \sqrt{\frac{L}{9{\cdot}8}}$ (1)<br>$5{\cdot}7 \times 10^{-2} = \frac{L}{9{\cdot}8}$<br>$L = 5{\cdot}6\,\text{m}$ (1) | 2 |
| | (e) | | Amplitude decreases | 1 |
| 8. | (a) | | Laser Grating Screen<br>labels (1)<br>order (1) | 2 |
| | (b) | | m = 1<br>Δx<br>m = 0<br>θ<br>m = −1<br>D (1)<br>$\Delta x$ and $D$ are measured for each fringe (1)<br>$\theta$ calculated using<br>$\tan\theta = \frac{\Delta x}{D}$ (1)<br>Note, for $m = -1$ the sin will be negative (below the axis). | 3 |
| | (c) | | $d\sin\theta = m\lambda$<br>$\sin\theta = \frac{\lambda}{d}m$<br>Gradient $= \frac{\lambda}{d}$ (1)<br>Gradient $= 0{\cdot}202$ (1)<br>$d = \frac{1}{300}\,\text{lines mm}^{-1}$<br>$d = 3{\cdot}3 \times 10^{-6}\,\text{lines m}^{-1}$ (1)<br>$\lambda = \text{gradient} \times d$<br>$= 0{\cdot}202 \times 3{\cdot}3 \times 10^{-6}\,\text{m}$<br>$\lambda = 6{\cdot}67 \times 10^{-7}\,\text{m}$ (1) | 4 |
| | (d) | | Percentage uncertainty in gradient $= \pm\left(\frac{0{\cdot}0067}{0{\cdot}202} \times 100\right)$<br>$= \pm 3{\cdot}3\%$ (1)<br>Absolute uncertainty<br>$= \pm\left(\frac{3{\cdot}3}{100} \times 6{\cdot}67 \times 10^{-7}\right)$ (1)<br>$= \pm 0{\cdot}2 \times 10^{-7}\,\text{m}$ (1) | 3 |

| Question | | | Expected response | Max mark |
|---|---|---|---|---|
| | (e) | | **Any two of the following:**<br>• Repeat readings<br>• Increase distance between grating and screen<br>• Calculate absolute uncertainty in $\sin\theta$<br>• Estimate scale reading, calibration uncertainty in the measured distances<br>• Combine uncertainties in corresponding measured distances<br>• Plot error bars on graph for $\sin\theta$ | 2 |
| 9. | (a) | | $\lambda = 2 \times 0{\cdot}15$ **(1)**<br>$v = f\lambda$ **(1)**<br>$v = 250 \times 0{\cdot}3$<br>$(v = 75\,\text{m}\,\text{s}^{-1})$ **(1)** | 3 |
| | (b) | | Percentage uncertainty in<br>$\lambda = \pm\left(\frac{0{\cdot}005 \times 100}{0{\cdot}150}\right)$<br>$= \pm 3{\cdot}3\%$ **(1)**<br>Percentage uncertainty in<br>$f = \pm\left(\frac{10 \times 100}{250}\right)$<br>$= \pm 4\%$ **(1)**<br>Percentage uncertainty in<br>$v = \pm\sqrt{(3{\cdot}3^2 + 4^2)}$<br>$= \pm 5\%$ **(1)**<br>Absolute uncertainty<br>$= \pm\left(\frac{75 \times 5}{100}\right)$ **(1)**<br>$= \pm 4\,\text{m}\,\text{s}^{-1}$<br>$v = (75 \pm 4)\,\text{m}\,\text{s}^{-1}$ **(1)** | 5 |
| | (c) | (i) | % uncertainty in $\lambda$ will increase | 1 |
| | | (ii) | Measure the distance over several nodes and take the mean value. | 1 |
| 10. | | | **Open-ended question**<br>Answers might include:<br>• He was incorrect.<br>• Many mysteries still in Physics.<br>• Research continues in numerous fields – can go on to mention any specific or list numerous discoveries.<br>• Might concentrate on one particular aspect – space travel, satellites, communication, computers, weapons, nuclear power, radar, GPS systems, particle physics, CERN, cosmology, special and general relativity, dark matter, dark energy, Cern – Higgs Boson, medical physics, quantum physics, the list is endless ...<br>• The main thing is really that no matter the age, research will continue to push back the boundary of science and understand more about our origin and how the universe works. | 3 |
| 11. | (a) | (i) | [diagram of field lines around two negative charges]<br>For shape **(1)**<br>For direction **(1)** | 2 |

| Question | | | Expected response | Max mark |
|---|---|---|---|---|
| | | (ii) | $V = \frac{Q}{4\pi\varepsilon_0 r}$ **(1)**<br>$V_X = V_{Q1} + V_{Q2}$<br>$V_X = \frac{-4\times10^{-6}}{4\pi\times8{\cdot}85\times10^{-12}\times0{\cdot}3} + \frac{-4\times10^{-6}}{4\pi\times8{\cdot}85\times10^{-12}\times0{\cdot}3}$ **(1)**<br>$V_X = -2{\cdot}4\times10^5\,\text{V}$ **(1)** | 3 |
| | (b) | (i) | −8·0 μC<br>Q3<br>0·50 m<br>0·40 m<br>−4·0 μC<br>θ<br>Q1<br>0·30 m **(1)**<br>$r = 0{\cdot}5\,\text{m}$ **(1)**<br>$F = \frac{Qq}{4\pi\varepsilon_0 r^2}$ **(1)**<br>$F = \frac{-8\times10^{-6}\times-4\times10^{-6}}{4\pi\times8{\cdot}85\times10^{-12}\times0{\cdot}5^2}$<br>$F = 1{\cdot}2\,\text{N}$ | 3 |
| | | (ii) | $F = 2\times1{\cdot}2\cos37^\circ$ **(1)**<br>$F = 1{\cdot}9\,\text{N}$ **(1)**<br>Direction $(000)^\circ$ **(1)** | 3 |
| **12.** | (a) | | $2{\cdot}4\times10^3\,\text{eV}$<br>$= 1{\cdot}60\times10^{-19}\times2400\,\text{J}$<br>$= \frac{1}{2}\times m\times v^2$ **(1)**<br>$1{\cdot}60\times10^{-19}\times2400$<br>$= \frac{1}{2}\times9{\cdot}11\times10^{-31}\times v^2$ **(1)**<br>$v = 2{\cdot}90\times10^7\,\text{m s}^{-1}$ **(1)** | 3 |
| | (b) | (i) | Constant force/acceleration in **vertical direction** **(1)**<br>Constant speed in horizontal direction **(1)** | 2 |
| | | (ii) | No unbalanced force acting on electron | 1 |

| Question | | | Expected response | Max mark |
|---|---|---|---|---|
| | (c) | (i) | $E = \frac{V}{d}$<br>$E = \frac{100}{0{\cdot}01}$ **(1)**<br>$E = (10^4\,\text{Vm}^{-1})$<br>$F = QE$<br>$F = 1{\cdot}60\times10^{-19}\times\frac{100}{0{\cdot}01}$ **(1)**<br>$F = 1{\cdot}60\times10^{-15}\,\text{N}$<br>$a = \frac{F}{m}$<br>$a = \frac{1{\cdot}60\times10^{-15}}{9{\cdot}11\times10^{-31}}$ **(1)**<br>$a = 1{\cdot}76\times10^{15}\,\text{m s}^{-2}$ | 3 |
| | | (ii) | $t = \frac{s_H}{v_H}$<br>$t = 5{\cdot}17\times10^{-10}\,\text{s}$ **(1)**<br>$v_v = u_v + a_v t$ **(1)**<br>$v_v = 0 + 1{\cdot}76\times10^{15} \times 5{\cdot}17\times10^{-10}$ **(1)**<br>$v_v = 9{\cdot}10\times10^5\,\text{m s}^{-1}$ **(1)** | 4 |
| | (d) | | Length scanned decreases **(1)**<br>$v_H$ increases/greater acceleration **(1)**<br>Shorter time between plates<br>**OR**<br>Vertical speed is less on leaving the plates **(1)** | 3 |
| **13.** | (a) | | 0·37<br>$0{\cdot}37\times12 = 4{\cdot}44\,\text{V}$<br>Reading 4·44 V from graph (accept 4·4 – 4·5 V) **(1)**<br>This gives 9·5 ms from graph **(1)** | 2 |
| | (b) | | $t = RC$ **(1)**<br>$9{\cdot}5\times10^{-3} = R\times385\times10^{-6}$ **(1)**<br>$R = 25\,\Omega$ **(1)** | 3 |

## ADVANCED HIGHER PHYSICS 2016

| Question | | | Answer | Max mark |
|---|---|---|---|---|
| 1. | (a) | | $v = 0 \cdot 135t^2 + 1 \cdot 26t$<br>$a = \frac{dv}{dt} = 0 \cdot 135 \times 2t + 1 \cdot 26$ **1**<br>$a = (0 \cdot 135 \times 2 \times 15 \cdot 0) + 1 \cdot 26$ **1**<br>$a = 5 \cdot 31\ \text{m s}^{-2}$ **1** | 3 |
| | (b) | | $v = 0 \cdot 135t^2 + 1 \cdot 26t$<br>$s = \int_0^{15 \cdot 0} v.dt = [0 \cdot 045t^3 + 0 \cdot 63t^2]_0^{15 \cdot 0}$ **1**<br>$s = (0 \cdot 045 \times 15 \cdot 0^3) + (0 \cdot 63 \times 15 \cdot 0^2)$ **1**<br>$s = 294$ m **1** | 3 |
| 2. | (a) | (i) | Velocity changing<br>Or changing direction<br>Or an unbalanced force is acting<br>Or a centripetal/central/radial force is acting | 1 |
| | | (ii) | Towards the centre | 1 |
| | (b) | (i) (A) | SHOW QUESTION<br>$\omega = \frac{d\theta}{dt}$ OR $\omega = \frac{\theta}{t}$ **1**<br>$\omega = \frac{1 \cdot 5 \times 2\pi}{2 \cdot 69}$ **1**<br>$\omega = 3 \cdot 5\ \text{rad s}^{-1}$ | 2 |
| | | (B) | $F = mr\omega^2$ **1**<br>$F = 0 \cdot 059 \times 0 \cdot 48 \times 3 \cdot 5^2$ **1**<br>$F = 0 \cdot 35$ N **1** | 3 |
| | | (C) | $W = mg$<br>$W = 0 \cdot 059 \times 9 \cdot 8$ **1**<br>$T^2 = 0 \cdot 35^2 + (0 \cdot 059 \times 9 \cdot 8)^2$ **1**<br>$T = 0 \cdot 68$ N **1** | 3 |
| | | (ii) | In a straight line at a tangent to the circle | 1 |

| Question | | | Answer | Max mark |
|---|---|---|---|---|
| 3. | (a) | | $v = \sqrt{\frac{2GM}{r}}$ **1**<br>$v = \sqrt{\frac{2 \times 6 \cdot 67 \times 10^{-11} \times 9 \cdot 5 \times 10^{12}}{2 \cdot 1 \times 10^3}}$ **1**<br>$v = 0 \cdot 78\ (\text{m s}^{-1})$ **1**<br>(Lander returns to surface as) lander v less than escape velocity of comet **1** | 4 |
| | (b) | (i) | SHOW QUESTION<br>$(F_g = W)$<br>$\frac{GMm}{r^2} = mg$ **1** for both eqns, **1** for equating<br>$g = \frac{GM}{r^2}$<br>$g = \frac{6 \cdot 67 \times 10^{-11} \times 9 \cdot 5 \times 10^{12}}{(2 \cdot 1 \times 10^3)^2}$ **1**<br>$g = 1 \cdot 4 \times 10^{-4}\ \text{N kg}^{-1}$ | 3 |
| | | (ii) | Height will be greater **1**<br>Because '$a$' reduces **1**<br>with height **1** | 3 |
| 4. | (a) | | $b = \frac{L}{4\pi r^2}$ **1**<br>$1 \cdot 05 \times 10^{-9} = \frac{L}{4\pi (9 \cdot 94 \times 10^{16})^2}$ **1**<br>$L = 1 \cdot 30 \times 10^{26}$ W **1** | 3 |
| | (b) | | $L = 4\pi r^2 \sigma T^4$ **1**<br>$1 \cdot 30 \times 10^{26} = 4\pi (5 \cdot 10 \times 10^8)^2$<br>$\times 5 \cdot 67 \times 10^{-8} \times T^4$ **1**<br>$T = 5150$ K **1** | 3 |
| | (c) | | That the star is a black body (emitter/radiator)<br>**OR**<br>The star is spherical/constant radius<br>**OR**<br>The surface temperature of the star is constant/uniform<br>**OR**<br>No energy absorbed between star and Earth | 1 |

| Question | | | Answer | Max mark |
|---|---|---|---|---|
| **5.** | (a) | (i) | Frames of reference that are accelerating (with respect to an inertial frame) | 1 |
| | | (ii) | It is impossible to tell the difference between the effects of gravity and acceleration. | 1 |
| | (b) | (i) | Any convex upward parabola | 1 |
| | | (ii) | Any straight line | 1 |
| | (c) | | The clock on the surface of the Earth would run more slowly. 1<br>The (effective) gravitational field for the spacecraft is smaller. 1<br>**OR** vice versa | 2 |
| **6.** | | | This is an open-ended question.<br>Demonstrates no understanding 0 marks<br>Demonstrates limited understanding 1 marks<br>Demonstrates reasonable understanding 2 marks<br>Demonstrates good understanding 3 marks<br>**1 mark:** The student has demonstrated a limited understanding of the physics involved. The student has made some statement(s) which is/are relevant to the situation, showing that at least a little of the physics within the problem is understood.<br>**2 marks:** The student has demonstrated a reasonable understanding of the physics involved. The student makes some statement(s) which is/are relevant to the situation, showing that the problem is understood.<br>**3 marks:** The maximum available mark would be awarded to a student who has demonstrated a good understanding of the physics involved. The student shows a good comprehension of the physics of the situation and has provided a logically correct answer to the question posed. This type of response might include a statement of the principles involved, a relationship or an equation, and the application of these to respond to the problem. This does not mean the answer has to be what might be termed an "excellent" answer or a "complete" one. | 3 |
| **7.** | (a) | (i) | $T_K = 15 + 273$ 1<br>$T_{kelvin} = \frac{b}{\lambda_{peak}}$<br>$288 = \frac{2 \cdot 89x10^{-3}}{\lambda_{peak}}$ 1<br>$\lambda_{peak} = 1 \cdot 0 \times 10^{-5}$ m 1 | 3 |
| | | (ii) | Infrared | 1 |
| | (b) | (i) | (Curve) A 1<br>Peak at shorter wavelength/higher frequency (as temperature is higher) 1<br>**OR**<br>Higher/greater (peak) intensity (as greater energy) 1 | 2 |

| Question | | | Answer | Max mark |
|---|---|---|---|---|
| | | (ii) | Curve asymptotic to y-axis and decreasing with increased wavelength | 1 |
| **8.** | (a) | (i) | The uncertainty in the momentum (in the x-direction) | 1 |
| | | (ii) | The precise position of a particle/system and its momentum cannot both be known at the same instant. **1**<br>**OR**<br>If the uncertainty in the location of the particle is reduced, the minimum uncertainty in the momentum of the particle will increase (or vice-versa). **1**<br>**OR**<br>The precise energy and lifetime of a particle cannot both be known at the same instant. **1**<br>**OR**<br>If the uncertainty in the energy of the particle is reduced, the minimum uncertainty in the lifetime of the particle will increase (or vice-versa). **1** | 1 |
| | (b) | (i) | $\lambda = \frac{h}{p}$ **1**<br>$\lambda = \frac{6 \cdot 63 \times 10^{-34}}{6 \cdot 5 \times 10^{-24}}$ **1**<br>$\lambda = 1 \cdot 0 \times 10^{-10}$ (m) **1**<br>slit width $0 \cdot 1$ nm used **1** | 4 |
| | | (ii) | $\Delta x \, \Delta p_x \geq \frac{h}{4\pi}$ **1**<br>$\Delta x \times 6 \cdot 5 \times 10^{-26} \geq \frac{6 \cdot 63 \times 10^{-34}}{4\pi}$ **1**<br>$\Delta x \geq 8 \cdot 1 \times 10^{-10}$<br>min uncertainty<br>$= 8 \cdot 1 \times 10^{-10}$ m **1** | 3 |

| Question | | | Answer | Max mark |
|---|---|---|---|---|
| | (b) | (iii) | Electron behaves like a wave<br>"Electron shows interference"<br>Uncertainty in position is greater than slit separation<br>Electron passes through both slits<br>Any three of the statements can be awarded 1 mark each | **3** |
| **9.** | (a) | | SHOW QUESTION<br>$m\frac{v^2}{r} = Bqv$<br>1 for both relationships<br>1 for equating<br>$r = \frac{mv}{Bq}$ | **2** |
| | (b) | (i) | $1 \cdot 50$ (MeV)<br>$= 1 \cdot 50 \times 10^6$<br>$\times 1 \cdot 60 \times 10^{-19}$<br>$= 2 \cdot 40 \times 10^{-13}$ (J) **1**<br>$E_k = \frac{1}{2} mv^2$ **1**<br>$2 \cdot 40 \times 10^{-13} = 0 \cdot 5 \times 3 \cdot 34$<br>$\times 10^{-27} \times v^2$ **1**<br>$v = 1 \cdot 20 \times 10^7$ ms$^{-1}$ **1** | **4** |
| | | (ii) | $r = \frac{mv}{Bq}$<br>$2 \cdot 50 = \frac{3 \cdot 34 \times 10^{-27} \times 1 \cdot 20 \times 10^7}{B \times 1 \cdot 60 \times 10^{-19}}$ **1**<br>$B = 0 \cdot 100$ T **1** | **2** |
| | | (iii) | r will be less **1**<br>$r \propto \frac{m}{q}$<br>**and**<br>q increases more than m does or q doubles but m increases by a factor of 1·5 **1** | **2** |
| **10.** | (a) | (i) | Displacement is proportional to and in the opposite direction to the acceleration. | **1** |

| Question | | | Answer | Max mark |
|---|---|---|---|---|
| | | (ii) | SHOW QUESTION<br>$y = A\cos \omega t$<br>$\frac{dy}{dt} = -\omega A\sin \omega t$<br>$\frac{d^2y}{dt^2} = -\omega^2 A\cos \omega t$ 1<br>$\frac{d^2y}{dt^2} = -\omega^2 y$ 1<br>$\frac{d^2y}{dt^2} + \omega^2 y = 0$ | 2 |
| | (b) | (i) | SHOW QUESTION<br>$T = \frac{12 \cdot 0}{10}$ 1<br>$\omega = \frac{2\pi}{T}$ 1<br>$\omega = \frac{2\pi \times 10}{12}$ 1<br>$\omega = 5 \cdot 2$ rad $s^{-1}$ | 3 |
| | | (ii) | $v = (\pm)\omega\sqrt{A^2 - y^2}$ 1<br>$v = 5 \cdot 2 \times 0 \cdot 04$ 1<br>$v = 0 \cdot 21$ m $s^{-1}$ 1 | 3 |
| | | (iii) | $E_P = \frac{1}{2} m\omega^2 y^2$ 1<br>$E_P = \frac{1}{2} \times 1 \cdot 5 \times 5 \cdot 2^2 \times 0 \cdot 04^2$ 1<br>$E_P = 0 \cdot 032$ J 1 | 3 |
| | (c) | (i) | Any valid method of damping | 1 |
| | | (ii) | Amplitude of harmonic wave reducing | 1 |
| 11. | (a) | | $\frac{1}{\lambda} = 0 \cdot 357$ 1<br>$\lambda = \frac{1}{0 \cdot 357}$<br>$v = f\lambda$ 1<br>$v = 118 \times \frac{1}{0 \cdot 357}$ 1<br>$v = 331$ m $s^{-1}$ 1 | 4 |
| | (b) | | $E = kA^2$ 1<br>$\frac{E_1}{A_1^2} = \frac{E_2}{A_2^2}$<br>$\frac{1}{0 \cdot 250^2} = \frac{0 \cdot 5}{A_2^2}$ 1<br>$A_2 = 0 \cdot 177$ (m) 1<br>$\Delta y = 0 \cdot 177 \sin 2\pi (118t + 0 \cdot 357x)$ 1 | 4 |
| 12. | (a) | | (The axes should be arranged) at 90° to each other (e.g. horizontal and vertical). | 1 |
| | (b) | | The filter for each eye will allow light from one projected image to pass through 1<br>while blocking the light from the other projector. 1 | 2 |
| | (c) | | There will be no change to the brightness. 1<br>Light from the lamp is unpolarised. 1 | 2 |
| | (d) | | (As the student rotates the filter,) the image from one projector will decrease in brightness, while the image from the other projector will increase in brightness. (The two images are almost identical.) 1 | 1 |
| 13. | (a) | | SHOW QUESTION<br>$V = \frac{1}{4\pi\varepsilon_o} \frac{Q_1}{r}$ 1<br>$V = \frac{1}{4\pi \times 8 \cdot 85 \times 10^{-12}} \frac{12 \times 10^{-9}}{0 \cdot 30}$ 1<br>$V = (+)360$ V | 2 |

| Question | | | Answer | Max mark |
|---|---|---|---|---|
| | (b) | (i) | $V = -360$ (V) **1**<br>$V = \frac{1}{4\pi\varepsilon_o}\frac{Q_2}{r}$<br>$-360 = \frac{Q_2}{4\pi \times 8\cdot85 \times 10^{-12} \times 0\cdot40}$ **1**<br>$Q_2 = -1\cdot6 \times 10^{-8}$ C **1** | 3 |
| | | (ii) | $E_1 = \frac{1}{4\pi\varepsilon_o}\frac{Q_1}{r^2}$ **1**<br>$E_1 = \frac{1}{4\pi \times 8\cdot85 \times 10^{-12}}\frac{12 \times 10^{-9}}{0\cdot30^2}$ **1**<br>$E_1 = 1200$ (N $C^{-1}$ *to right*)<br>$E_2 = \frac{1}{4\pi \times 8\cdot85 \times 10^{-12}}\frac{1\cdot6 \times 10^{-8}}{0\cdot40^2}$ **1**<br>$E_2 = 900$ (N $C^{-1}$ to right)<br>*Total* $= 2100$ N $C^{-1}$ (to right) **1** | 4 |
| | | (iii) | Shape of attractive field, including correct direction **1**<br>Skew in correct direction **1** | 2 |
| **14.** | (a) | | $B = \frac{\mu_o I}{2\pi r}$ **1**<br>$B = 5 \times 10^{-6} = \frac{4\pi \times 10^{-7} \times I}{2\pi \times 0\cdot1}$ **1**<br>$I = 2\cdot5$ A **1** | 3 |
| | (b) | (i) | Ignore calibration (less than 1/3)<br>$\%u/c = \frac{0\cdot002}{0\cdot1} \times 100 = 2\%$ **1** | 1 |
| | | (ii) | Reading $5 = \frac{0\cdot1}{5} \times 100 = 2\%$ **1**<br>Total% = √(reading%$^2$+ calibration%$^2$) **1**<br>Total % = √$(1\cdot5^2 + 2^2)$<br>= 2·5% **1** | 3 |
| | | (iii) | Total % = √$(2^2 + 2\cdot5^2)$<br>= √10·25% **1**<br>abs u/c = $\frac{\sqrt{10\cdot25}}{100} \times 2\cdot5$ **1**<br>= 0·08 A | 2 |
| | (c) | | Uncertainty in measuring exact distance from wire to position of sensor | 1 |
| **15.** | (a) | (i) | $gradient = \frac{8\cdot3 \times 10^{-10}}{10^3}$<br>$= 8\cdot3 \times 10^{-13}$ **1**<br>-------------------------<br>$gradient = \varepsilon_0 A$<br>$8\cdot3 \times 10^{-13} = \varepsilon_0 \times 9\cdot0 \times 10^{-2}$ **1**<br>$\varepsilon_0 = 9\cdot2 \times 10^{-12}$ F $m^{-1}$ **1** | 3 |
| | | (ii) | $c = \frac{1}{\sqrt{\varepsilon_0\mu_0}}$ **1**<br>$c = \frac{1}{\sqrt{9\cdot2 \times 10^{-12} \times 4\pi \times 10^{-7}}}$ **1**<br>$c = 2\cdot9 \times 10^8$ m $s^{-1}$ **1** | 3 |
| | (b) | | Systematic uncertainty specific to capacitance or spacing measurement | 1 |
| **16.** | (a) | | $I = \frac{2}{5}mr^2$ **1**<br>$I = \frac{2}{5} \times 3\cdot8 \times 0\cdot053^2$ **1**<br>$I = 4\cdot3 \times 10^{-3}$ kg $m^2$ **1** | 3 |
| | (b) | (i) | Labelling & scales **1**<br>Plotting **1**<br>Best fit line **1** | 3 |
| | | (ii) | *gradient* = 1·73 or consistent with candidate's best fit line **1**<br>-------------------------<br>$2gh = \left(\frac{I}{mr^2} + 1\right)v^2$<br>$\frac{2gh}{v^2} = \left(\frac{I}{mr^2} + 1\right)$<br>$1\cdot73 = \left(\frac{I}{3\cdot8 \times 0\cdot053^2} + 1\right)$ **1**<br>$I = 7\cdot8 \times 10^{-3}$ kgm$^2$ **1** | 3 |

| Question | | | Answer | Max mark |
|---|---|---|---|---|
| | (c) | | This is an open-ended question.<br>Demonstrates no understanding 0 marks<br>Demonstrates limited understanding 1 marks<br>Demonstrates reasonable understanding 2 marks<br>Demonstrates good understanding 3 marks<br>**1 mark:** The student has demonstrated a limited understanding of the physics involved. The student has made some statement(s) which is/are relevant to the situation, showing that at least a little of the physics within the problem is understood.<br>**2 marks**: The student has demonstrated a reasonable understanding of the physics involved. The student makes some statement(s) which is/are relevant to the situation, showing that the problem is understood.<br>**3 marks:** The maximum available mark would be awarded to a student who has demonstrated a good understanding of the physics involved. The student sh ows a good comprehension of the physics of the situation and has provided a logically correct answer to the question posed. | 3 |
| | | | This type of response might include a statement of the principles involved, a relationship or an equation, and the application of these to respond to the problem. This does not mean the answer has to be what might be termed an "excellent" answer or a "complete" one. | |

# Acknowledgements

Permission has been sought from all relevant copyright holders and Hodder Gibson is grateful for the use of the following:

Image © Calvin Chan/Shutterstock.com (2016 page 3).

Hodder Gibson would like to thank SQA for use of any past exam questions that may have been used in model papers, whether amended or in original form.